Advanced
Genetic Analysis

Advanced Genetic Analysis

Ralph Phillip

Editor

KOROS PRESS LIMITED
London, UK

Advanced Genetic Analysis

© 2012
Printed in 2017 for Sale in the Indian Subcontinent

Published by
Koros Press Limited
3 The Pines, Rubery B45 9FF, Rednal,
Birmingham, United Kingdom

Tel.: +44-7826-930152
Email: info@korospress.com
www.korospress.com

ISBN: 978-1-78163-255-0

Editor: Ralph Phillip

Printed in UK

British Library Cataloguing in Publication Data
A CIP record for this book is available from the British Library

10 9 8 7 6 5 4 3 2 1

Exclusively distributed by CBS Publishers & Distributors Pvt. Ltd.
Sales & Distribution Rights only for India, Pakistan, Bangladesh, Sri Lanka, Nepal and Bhutan.This book is not to be sold outside these territories.

Contents

Preface　　　　　　　　　　　　　　　　　　　　　　　　　*vii*

1. Genetics, Heredity and Variations　　　　　　　　1

Genetic Linkage · Heredity · Gene Interaction · Making Sense
of the Complex · Genome · Sequencing and Mapping

2. Identification of Genetic Material　　　　　　　19

Direct Evidences for DNA as the Genetic Material · Mode of
Infection of Bacteriophages and Identification of DNA as their
Genetic Material · Indirect Evidences for DNA as the Genetic
Material · Base Pair · Nucleic Acid Analogues · Base Analogs
and Intercalators · Protein Synthesis · DNA and Protein Synthesis
· Regulation of Gene Expression or Regulation of Protein Synthesis
· Epigenetics · Modification of DNA · Circuitry · Eubacteria -
Gene Regulation and Protein Synthesis · Gene Expression
Regulates Cell Differentiation · Transcription Factors and
Transcriptional Control in Eukaryotic Cells · Nucleic Acids to
Amino Acids: DNA Specifies Protein · Myofilaments · The Z-line
· Slioing Mechanism · Energetics of Contraction · Sarcoplasmic
Reticulum · Cytogenetices · Karyotype

3. Genetic Effects of Cross-fertilization　　　　82

Random Mating · Diploid Chromosome Behaviour and Panmixis
· Autotetraploid Chromosome Behaviour and Panmixis · Genetic
Effects of Inbreeding · Diploid Chromosome Behaviour and
Inbreeding · Autotetraploid Chromosome Behaviour and Self-
Fertilization

4. Genetic Improvement of Plant　　　　　　　129

Introduction · Micropropagation of Bamboo · Monitoring Genetic
Stability in Bamboo Tissue Culture · Tissue Culture Techniques
· Micropropagation · Cultured Cells and Tissues · Plant Tissue
Culture · Pathways of Cultured Cells and Tissues

5. Genetic Resources and Plant Breeding　　　158

Uses of Marker-aided Selection · Types of Genetic Markers
· Locating Genes of Major Effect · Mapping Genes Controlling

Quantitative Traits · Alien Gene Transfer · Cloning of Plant Traits

6. Life's Genetic Tree   **193**

Everything Evolves · Living Laboratories · Plant Breeding · Domestication · Classical Plant Breeding · Modern Plant Breeding · Issues and Concerns

7. Quantitative Genetics, Genomics, Medical Genetics and Behavioural Genetics   **210**

Quantitative Genetics · Genomics · Medical Genetics · Behavioural Genetics · Genetic Diversity · Genetic Erosion · Genetic Hitchhiking · Genetic Monitoring · Genetic Pollution · Population Genetics · Genetic Analysis of Adenohypophysis Formation in Zebrafish · Dual Inheritance Theory

Bibliography   273

Index   277

Preface

Genetics, a discipline of biology, is the science of genes, heredity, and variation in living organisms. Genetics deals with the molecular structure and function of genes, with gene behaviour in the context of a cell or organism with patterns of inheritance from parent to offspring, and with gene distribution, variation and change in populations. Given that genes are universal to living organisms, genetics can be applied to the study of all living systems, from viruses and bacteria, through plants, to humans, as in medical genetics. The fact that living things inherit traits from their parents has been used since prehistoric times to improve crop plants and animals through selective breeding. However, the modern science of genetics, which seeks to understand the process of inheritance, only began with the work of Gregor Mendel in the mid-19th century. Although he did not know the physical basis for heredity, Mendel observed that organisms inherit traits via discrete units of inheritance, which are now called genes. Genes correspond to regions within DNA, a molecule composed of a chain of four different types of nucleotides—the sequence of these nucleotides is the genetic information organisms inherit. DNA naturally occurs in a double stranded form, with nucleotides on each strand complementary to each other. Each strand can act as a template for creating a new partner strand. This is the physical method for making copies of genes that can be inherited.

The sequence of nucleotides in a gene is translated by cells to produce a chain of amino acids, creating proteins—the order of amino acids in a protein corresponds to the order of nucleotides in the gene. This relationship between nucleotide sequence and amino acid sequence is known as the genetic code. The amino acids in a protein determine how it folds into a three-dimensional shape; this structure is, in turn, responsible for the protein's function. Proteins carry out almost all the functions needed for cells to live. A change to the DNA in a gene can change a protein's amino acds, changing its shape and function: this can have a dramatic effect in the cell and on the organism as a whole. Although genetics plays a large role in the appearance and behaviour of organisms, it is the combination of genetics with what an organism experiences that determines the ultimate outcome. For example, while genes play a role in determining an organism's size, the nutrition and health it experiences after inception also have a large effect.

Although the science of genetics began with the applied and theoretical work of Gregor Mendel in the mid-19th century, other theories of inheritance preceded Mendel. A popular theory during Mendel's time was the concept of blending inheritance: the idea that individuals inherit a smooth blend of traits from their parents. Mendel's work disproved this, showing that traits are composed of combinations of distinct genes rather than a continuous blend. Another theory that had some support at that time was the inheritance of acquired characteristics: the belief that individuals inherit traits strengthened by their parents. This theory is now known to be wrong—the experiences of individuals do not affect the genes they pass to their children. Other theories included the pangenesis of Charles Darwin and Francis Galton's reformulation of pangenesis as both particulate and inherited.

This book is a comprehensive and analytical study of Molecular Genetics which covers, nearly, all important aspects of DNA and molecular testing.

—Editor

Chapter 1

Genetics, Heredity and Variations

Genetic Linkage

Genetic linkage is a term which describes the tendency of certain loci or alleles to be inherited together. Genetic loci on the same chromosome are physically close to one another and tend to stay together during meiosis, and are thus genetically *linked*.

Background

At the beginning of normal meiosis, a chromosome pair (made up of a chromosome from the mother and a chromosome from the father) intertwine and exchange sections or fragments of chromosome. The pair then breaks apart to form two chromosomes with a new combination of genes that differs from the combination supplied by the parents. Through this process of recombining genes, organisms can produce offspring with new combinations of maternal and paternal traits that may contribute to or enhance survival.

This crossing over of DNA can cause alleles previously on the same chromosome to be separated and end up in different daughter cells. The further the two alleles are apart, the greater the chance that a cross-over event may occur between them, possibly separating the alleles.

The relative distance between two genes can be calculated using the offspring of an organism showing two linked genetic traits, and finding the percentage of the offspring where the two traits do not run together. The higher the percentage of descendants that does not show both traits, the farther apart on the chromosome the two genes are.

Among individuals of an experimental population or species, some phenotypes or traits occur randomly with respect to one another in a manner known as independent assortment. Today scientists understand that independent assortment occurs when the genes affecting the phenotypes are found on different chromosomes or separated by a great enough distance on the same chromosome that recombination occurs at least half of the time.

An exception to independent assortment develops when genes appear near one another on the same chromosome. When genes occur on the same chromosome, they are usually inherited as a single unit. Genes inherited in this way are said to be linked, and are referred to as "linkage groups." For example, in fruit flies the genes affecting eye colour and wing length are inherited together because they appear on the same chromosome.

Genetic linkage was first discovered by the British geneticists William Bateson and Reginald Punnett shortly after Mendel's laws were rediscovered. The understanding of genetic linkage was expanded by the work of Thomas Hunt Morgan. Morgan's observation that the amount of crossing over between linked genes differs led to the idea that crossover frequency might indicate the distance separating genes on the chromosome.

Alfred Sturtevant, a student of Morgan's, first developed genetic maps, also known as linkage maps. Sturtevant proposed that the greater the distance between linked genes, the greater the chance that non-sister chromatids would cross over in the region between the genes. By working out the number of recombinants it is possible to obtain a measure for the distance between the genes. This distance is called a genetic map unit (m.u.), or a centimorgan and is defined as the distance between genes for which one product of meiosis in 100 is recombinant. A recombinant frequency (RF) of 1 % is equivalent to 1 m.u. But this equivalence is only a good approximate for small percentages; the largest percentage of recombinants cannot exceed 50%, which would be the situation where the two genes are at the extreme opposite ends of the same chromosomes. In this situation, any crossover events would result in an exchange of genes, but only an odd number of crossover events (a 50-50 chance between even and odd number of crossover events) would result in a recombinant product of meiotic crossover. A statistical interpretation of this is through the Haldane mapping function or the Kosambi mapping function, among others. A linkage map is created by finding the map distances between a number of traits that are present on the same chromosome, ideally avoiding having significant gaps between traits to avoid the inaccuracies that will occur due to the possibility of multiple recombination events.

Linkage Map

A linkage map is a genetic map of a species or experimental population that shows the position of its known genes or genetic markers relative to each other in terms of recombination frequency, rather than as specific physical distance along each chromosome. Linkage mapping is critical for identifying the location of genes that cause genetic diseases.

A genetic map is a map based on the frequencies of recombination between markers during crossover of homologous chromosomes. The greater the frequency of recombination (segregation) between two genetic markers, the farther apart they are assumed to be. Conversely, the lower the frequency of recombination between the markers, the smaller the physical distance between them. Historically, the markers originally used were detectable phenotypes (enzyme production, eye colour) derived from coding DNA sequences; eventually, confirmed or assumed noncoding DNA sequences such as microsatellites or those generating restriction fragment length polymorphisms (RFLPs) have been used.

Genetic maps help researchers to locate other markers, such as other genes by testing for genetic linkage of the already known markers.

A genetic map is not a physical map (such as a radiation reduced hybrid map) or gene map.

The LOD score (logarithm (base 10) of odds) is a statistical test often used for linkage analysis in human populations, and also in animal and plant populations. The LOD score compares the likelihood of obtaining the test data if the two loci are indeed linked, to the likelihood of observing the same data purely by chance. Positive LOD score favour the presence of linkage, whereas negative LOD scores indicate that linkage is less likely. The test was developed by Newton E. Morton. Computerized LOD score analysis is a simple way to analyze complex family pedigrees in order to determine the linkage between Mendelian traits (or between a trait and a marker, or two markers).

The method is described in greater detail by Strachan and Read. Briefly, it works as follows:

1. Establish a pedigree
2. Make a number of estimates of recombination frequency
3. Calculate a LOD score for each estimate

4. The estimate with the highest LOD score will be considered the best estimate.

The LOD score is calculated as follows:

NR denotes the number of non-recombinant offspring, and R denotes the number of recombinant offspring. The reason 0.5 is used in the denominator is that any alleles that are completely unlinked (e.g. alleles on separate chromosomes) have a 50% chance of recombination, due to independent assortment.

Theta is the recombinant fraction, it is equal to R/(NR + R).

In practice, LOD scores are looked up in a table which lists LOD scores for various standard pedigrees and various values of recombination frequency.

By convention, a LOD score greater than 3.0 is considered evidence for linkage. A LOD score of +3 indicates 1000 to 1 odds that the linkage being observed did not occur by chance. On the other hand, a LOD score less than-2.0 is considered evidence to exclude linkage. Although it is very unlikely that a LOD score of 3 would be obtained from a single pedigree, the mathematical properties of the test allow data from a number of pedigrees to be combined by summing the LOD scores. It is important to keep in mind that this traditional cutoff of LOD>+3 is an arbitrary one and that the difference between certain types of linkage studies, particularly analyses of complex genetic traits with hundreds of markers, these criteria should probably be modified to a somewhat higher cutoff.

Recombination frequency is the frequency that a single chromosomal crossover will take place between two genes during meiosis. Double crossovers would turn into no recombination. In this case we cannot tell if crossovers took place. If the loci we're analysing are very close (less than 7 cM) a double crossover is very unlikely. When distances become higher, the likelihood of a double crossover increases. Recombination frequency is a measure of genetic linkage and is used in the creation of a genetic linkage map. A centimorgan (cM) is a unit that describes a recombination frequency of 1%. In this way we can measure the genetic distance between two loci, based upon their recombination frequency. This is a good estimate of the real distance. As the likelihood of a double crossover increases we systematically underestimate the genetic distance between two loci.

During meiosis, chromosomes assort randomly into gametes, such that the segregation of alleles of one gene is independent of alleles of another gene. This is stated in Mendel's Second Law and is known as the law of independent assortment. The law of independent assortment always holds true for genes that are located on different chromosomes, but for genes that are on the same chromosome, it does not always hold true. As an example of independent assortment, consider the crossing of the pure-bred homozygote parental strain with genotype *AABB* with a different pure-bred strain with genotype *aabb*. A and a and B and b represent the alleles of genes A and B. Crossing these homozygous parental strains will result in F1 generation offspring with genotype AaBb.

The F1 offspring AaBb produces gametes that are *AB*, *Ab*, *aB*, and *ab* with equal frequencies (25%) because the alleles of gene A assort independently of the alleles for gene B during meiosis. Note that 2 of the 4 gametes (50 %)—*Ab* and *aB*—were not present in the parental generation. These gametes represent recombinant gametes. Recombinant gametes are those gametes that differ from both of the haploid gametes that made up the diploid cell. In this example, the recombination frequency is 50% since 2 of the 4 gametes were recombinant gametes.

The recombination frequency will be 50% when two genes are located on different chromosomes or when they are widely separated on the same chromosome. This is a consequence of independent assortment.

When two genes are close together on the same chromosome, they do not assort independently and are said to be linked. Whereas genes located on different chromosomes assort independently and have a recombination frequency of 50%, linked genes have a recombination frequency that is less than 50%.

As an example of linkage, consider the classic experiment by William Bateson and Reginald Punnett. They were interested in trait inheritance in the sweet pea and were studying two genes—the gene for flower colour (*P*, purple, and *p*, red) and the gene affecting the shape of pollen grains (*L*, long, and *l*, round). They crossed the pure lines *PPLL* and *ppll* and then self-crossed the resulting *PpLl* lines. According to Mendelian genetics, the expected phenotypes would occur in a 9:3:3:1 ratio of PL:Pl:pL:pl. To their surprise, they observed an increased frequency of PL and pl and a decreased frequency of Pl and pL.

Their experiment revealed linkage between the *P* and *L* alleles and the *p* and *l* alleles. The frequency of *P* occurring together with

L and with p occurring together with l is greater than that of the recombinant Pl and pL. The recombination frequency cannot be computed directly from this experiment, but intuitively it is less than 50%.

The progeny in this case received two dominant alleles linked on one chromosome (referred to as coupling or cis arrangement). However, after crossover, some progeny could have received one parental chromosome with a dominant allele for one trait (eg Purple) linked to a recessive allele for a second trait (eg round) with the opposite being true for the other parental chromosome (eg red and Long). This is referred to as repulsion or a trans arrangement. The phenotype here would still be purple and long but a test cross of this individual with the recessive parent would produce progeny with much greater proportion of the two crossover phenotypes. While such a problem may not seem likely from this example, unfavourable repulsion linkages do appear when breeding for disease resistance in some crops.

When two genes are located on the same chromosome, the chance of a crossover producing recombination between the genes is related to the distance between the two genes. Thus, the use of recombination frequencies has been used to develop linkage maps or genetic maps.

However, it is important to note that recombination frequency tends to underestimate the distance between two linked genes. This is because as the two genes are located further apart, the chance of double or even number of crossovers between them also increases. Double or even number of crossovers between the two genes results in them being cosegregated to the same gamete, yielding a parental progeny instead of the expected recombinant progeny.

The diploid nature of chromosomes allows for genes on different chromosomes to assort independently during sexual reproduction, recombining to form new combinations of genes. Genes on the same chromosome would theoretically never recombine, however, were it not for the process of chromosomal crossover. During crossover, chromosomes exchange stretches of DNA, effectively shuffling the gene alleles between the chromosomes. This process of chromosomal crossover generally occurs during meiosis, a series of cell divisions that creates haploid cells.

The probability of chromosomal crossover occurring between two given points on the chromosome is related to the distance between the points. For an arbitrarily long distance, the probability of crossover is high enough that the inheritance of the genes is effectively uncorrelated. For genes that are closer together, however, the lower probability of crossover means that the genes demonstrate genetic linkage—alleles for the two genes tend to be inherited together. The amounts of linkage between a series of genes can be combined to form a linear linkage map that roughly describes the arrangement of the genes along the chromosome.

Heredity

Heredity is the passing of traits to offspring (from its parent or ancestors). This is the process by which an offspring cell or organism acquires or becomes predisposed to the characteristics of its parent cell or organism. Through heredity, variations exhibited by individuals can accumulate and cause a species to evolve. The study of heredity in biology is called genetics, which includes the field of epigenetics.

Ancients had a variety of ideas about heredity: Theophrastus proposed that male flowers caused female flowers to ripen; Hippocrates speculated that "seeds" were produced by various body parts and transmitted to offspring at the time of conception, and Aristotle thought that male and female semen mixed at conception. Aeschylus, in 458 BC, proposed the male as the parent, with the female as a "nurse for the young life sown within her."

Various hereditary mechanisms were envisaged without being properly tested or quantified. These included blending inheritance and the inheritance of acquired traits. Nevertheless, people were able to develop domestic breeds of animals as well as crops through artificial selection. The inheritance of acquired traits also formed a part of early Lamarckian ideas on evolution.

In the 9th century AD, the Afro-Arab writer Al-Jahiz considered the effects of the environment on the likelihood of an animal to survive, and first described the struggle for existence. His ideas on the struggle for existence in the *Book of Animals* have been summarized as follows:

> *"Animals engage in a struggle for existence; for resources, to avoid being eaten and to breed. Environmental factors influence organisms to develop new characteristics to ensure survival, thus transforming into new species. Animals that survive to breed can pass on their successful characteristics to offspring."*

In 1000 AD, the Arab physician, Abu al-Qasim al-Zahrawi (known as Albucasis in the West), wrote the first clear description of haemophilia, a hereditary genetic disorder, in his *Al-Tasrif*. In this work, he wrote of an Andalusian family whose males died of bleeding after minor injuries.

During the 18th century, Dutch microscopist Antonie van Leeuwenhoek (1632–1723) discovered "animalcules" in the sperm of humans and other animals. Some scientists speculated they saw a "little man" (homunculus) inside each sperm. These scientists formed a school of thought known as the "spermists." They contended the only contributions of the female to the next generation were the womb in which the homunculus grew, and prenatal influences of the womb. An opposing school of thought, the ovists, believed that the future human was in the egg, and that sperm merely stimulated the growth of the egg. Ovists thought women carried eggs containing boy and girl children, and that the gender of the offspring was determined well before conception.

Pangenesis was an idea that males and females formed "pangenes" in every organ. These pangenes subsequently moved through their blood to the genitals and then to the children. The concept originated with the ancient Greeks, and influenced biology until as recently as a century ago. The terms "blood relative," "bloodline," "full-blooded," and "royal blood" are relics of pangenesis. Francis Galton, Charles Darwin's cousin, experimentally tested and disproved pangenesis during the 1870s.

Types of Heredity

An allele is said to be dominant if it is always expressed in the appearance of an organism (phenotype). For example, in peas the allele for green pods, G, is dominant to that for yellow pods, g. Since the allele for green pods is dominant, pea plants with the pair of alleles GG (homozygote) or Gg (heterozygote) will have green pods. The allele for yellow pods is recessive. The effects of this allele are only seen when it is present in both chromosomes, gg (homozygote).

The description of a mode of biological inheritance consists of three main categories:

1. Number of involved loci;
   - Monogenetic (also called "simple") – one locus
   - Oligogenetic – few loci
   - Polygenetic – many loci.

2. Involved chromosomes;
   - Autosomal – loci are not situated on a sex chromosome
   - Gonosomal – loci are situated on a sex chromosome
     — X-chromosomal – loci are situated on the X chromosome (the more common case)
     — Y-chromosomal – loci are situated on the Y chromosome
   - Mitochondrial – loci are situated on the mitochondrial DNA.
3. Correlation genotype–phenotype;
   - Dominant
   - Intermediate (also called "codominant")
   - Recessive.

These three categories are part of every exact description of a mode of inheritance in the above order. Additionally, more specifications may be added as follows:

4. Coincidental and environmental interactions;
   - Penetrance
     — Complete
     — Incomplete (percentual number)
   - Expressivity
     — Invariable
     — Variable
   - Heritability (in polygenetic and sometimes also in oligogenetic modes of inheritance)
   - Maternal or paternal imprinting phenomena.
5. Sex-linked interactions;
   - Sex-linked inheritance (gonosomal loci)
   - Sex-limited phenotype expression (e.g., cryptorchism)
   - Inheritance through the maternal line (in case of mitochondrial DNA loci)
   - Inheritance through the paternal line (in case of Y-chromosomal loci).
6. Locus–locus interactions;
   - Epistasis with other loci (e.g., overdominance)
   - Gene coupling with other loci

- Homozygotous lethal factors
- Semi-lethal factors.

Determination and description of a mode of inheritance is primarily achieved through statistical analysis of pedigree data. In case the involved loci are known, methods of molecular genetics can also be employed.

When Charles Darwin proposed his theory of evolution in 1859, one of its major problems was the lack of an underlying mechanism for heredity. Darwin believed in a mix of blending inheritance and the inheritance of acquired traits (pangenesis). Blending inheritance would lead to uniformity across populations in only a few generations and thus would remove variation from a population on which natural selection could act. This led to Darwin adopting some Lamarckian ideas in later editions of *On the Origin of Species* and his later biological works. Darwin's primary approach to heredity was to outline how it appeared to work (noticing that traits could be inherited which were not expressed explicitly in the parent at the time of reproduction, that certain traits could be sex-linked, etc.) rather than suggesting mechanisms.

Darwin's initial model of heredity was adopted by, and then heavily modified by, his cousin Francis Galton, who laid the framework for the biometric school of heredity. Galton rejected the aspects of Darwin's pangenesis model which relied on acquired traits.

The inheritance of acquired traits was shown to have little basis in the 1880s when August Weismann cut the tails off many generations of mice and found that their offspring continued to develop tails.

The idea of particulate inheritance of genes can be attributed to the Moravian monk Gregor Mendel who published his work on pea plants in 1865. However, his work was not widely known and was rediscovered in 1901. It was initially assumed the Mendelian inheritance only accounted for large (qualitative) differences, such as those seen by Mendel in his pea plants—and the idea of additive effect of (quantitative) genes was not realised until R.A. Fisher's (1918) paper, "The Correlation Between Relatives on the Supposition of Mendelian Inheritance."

In the 1930s, work by Fisher and others resulted in a combination of Mendelian and biometric schools into the modern evolutionary synthesis. The modern synthesis bridged the gap between experimental geneticists and naturalists; and between both and palaeontologists, stating that:

1. All evolutionary phenomena can be explained in a way consistent with known genetic mechanisms and the observational evidence of naturalists.

2. Evolution is gradual: small genetic changes, recombination ordered by natural selection. Discontinuities amongst species (or other taxa) are explained as originating gradually through geographical separation and extinction (not saltation).

3. Selection is overwhelmingly the main mechanism of change; even slight advantages are important when continued. The object of selection is the phenotype in its surrounding environment. The role of genetic drift is equivocal; though strongly supported initially by Dobzhansky, it was downgraded later as results from ecological genetics were obtained.

4. The primacy of population thinking: the genetic diversity carried in natural populations is a key factor in evolution. The strength of natural selection in the wild was greater than expected; the effect of ecological factors such as niche occupation and the significance of barriers to gene flow are all important.

5. In palaeontology, the ability to explain historical observations by extrapolation from micro to macro-evolution is proposed. Historical contingency means explanations at different levels may exist. Gradualism does not mean constant rate of change.

The idea that speciation occurs after populations are reproductively isolated has been much debated. In plants, polyploidy must be included in any view of speciation. Formulations such as 'evolution consists primarily of changes in the frequencies of alleles between one generation and another' were proposed rather later. The traditional view is that developmental biology ('evo-devo') played little part in the synthesis, but an account of Gavin de Beer's work by Stephen Jay Gould suggests he may be an exception.

Almost all aspects of the synthesis have been challenged at times, with varying degrees of success. There is no doubt, however, that the synthesis was a great landmark in evolutionary biology. It cleared up many confusions, and was directly responsible for stimulating a great deal of research in the post-World War II era.

Trofim Lysenko however caused a backlash of what is now called Lysenkoism in the Soviet Union when he emphasised Lamarckian ideas on the inheritance of acquired traits. This movement affected agricultural research and led to food shortages in the 1960s and seriously affected the USSR.

Gene Interaction

Gene–environment interaction (or genotype–environment interaction or GxE) is the phenotypic effect of interactions between genes and the environment.

Gene–environment interaction is exploited by plant and animal breeders to benefit agriculture. For example, plants can be bred to have tolerance for specific environments, such as high or low water availability. The way that trait expression varies across a range of environments for a given genotype is called its norm of reaction.

In genetic epidemiology it is frequently observed that diseases cluster in families, but family members may not inherit disease as such. Often, they inherit sensitivity to the effects of various environmental risk factors. Individuals may be differently affected by exposure to the same environment in medically significant ways. For example, sunlight exposure has a much stronger influence on skin cancer risk in fair-skinned humans than in individuals with an inherited tendency to darker skin.

Naive nature versus nurture debates assume that variation in a given trait is primarily due to either genetic variability or exposure to environmental experiences. The current scientific view is that neither genetics nor environment are solely responsible for producing individual variation, and that virtually all traits show gene–environment interaction. Evidence of statistical interaction between genetic and environmental risk factors is often used as evidence for the existence of an underlying mechanistic interaction.

In some combinations of environments and genotypic ranges, heritability can be 100% even while group differences are completely environmental. For heritability to be 100%, random variation in expression must not occur.

1. A classic example of gene–environment interaction is Tryon's artificial selection experiment on maze-running ability in rats. Tryon produced a remarkable difference in maze running ability in two selected lines after seven generations of selecting "bright" and "dull" lines by breeding the best and worst maze running rats with others of similar abilities. The difference between

these lines was clearly genetic since offspring of the two lines, raised under identical typical lab conditions, performed too differently. This difference disappeared in a single generation, if those rats were raised in an enriched environment with more objects to explore and more social interaction. This result shows that maze running ability is the product of a gene-by-environment interaction; the genetic effect is only seen under some environmental conditions.

2. Seven genetically distinct yarrow plants were collected and three cuttings taken from each plant. One cutting of each genotype was planted at low, medium, and high elevations, respectively. When the plants matured, no one genotype grew best at all altitudes, and at each altitude the seven genotypes fared differently. For example, one genotype grew the tallest at the medium elevation but attained only middling height at the other two elevations. The best growers at low and high elevation grew poorly at medium elevation. The medium altitude produced the worst overall results, but still yielded one tall and two medium-tall samples. Altitude had an effect on each genotype, but not to the same degree nor in the same way.

3. Phenylketonuria (PKU) is a human genetic condition caused by mutations to a gene coding for a particular liver enzyme. In the absence of this enzyme, an amino acid known as phenylalanine does not get converted into the next amino acid in a biochemical pathway, and therefore too much phenylalanine passes into the blood and other tissues. This disturbs brain development leading to mental retardation and other problems. PKU affects approximately 1 out of every 15,000 infants in the U.S. However, most affected infants do not grow up impaired because of a standard screening program used in the U.S. and other industrialized societies. Newborns found to have high levels of phenylalanine in their blood can be put on a special, phenylalanine-free diet. If they are put on this diet right away and stay on it, these children avoid the severe effects of PKU.

4. A functional polymorphism in the monoamine oxidase A (MAOA) gene promoter can moderate the association between early life trauma and increased risk for violence and antisocial behaviour. Low MAOA activity is a significant risk factor for aggressive and antisocial behaviour in adults who report

victimization as children. Persons who were abused as children but have a genotype conferring high levels of MAOA expression are less likely to develop symptoms of antisocial behaviour. These findings must be interpreted with caution, however, because gene association studies on complex traits are notorious for being very difficult to confirm.

- Doctors are interested in knowing whether disease can be prevented by reducing exposure to environmental risks. Gene–environment interaction means that some people carry genetic factors that confer susceptibility or resistance to a certain disorder in a particular environment. It has been argued that there may be significant public health benefits in using genetic information to stratify the allocation of environmental interventions that prevent disease, although this viewpoint is not universally held.

- Pharmacogenetics is the study of genetic variation that causes people to respond differently to drugs. The clinical importance of pharmacogenetics comes from the possibility that drug treatment can be made safer and more effective when the patient's genotype is known. Pharmacogenetic studies can be considered studies of gene–environment interaction, with drug treatment as the environmental variable.

Timing of Environmental Effects

The popular description of an animal being "born that way" does not necessarily discriminate genetic from environmental effects. In viviparous animals, such as humans, environmental influences may act during either pre-or post-natal development; similarly environmental influences may act before and after hatching to affect development in oviparous animals. Environmental influences in utero may be as strong and lasting as genetic or post-natal environmental influence.

Making Sense of the Complex

It is now becoming possible to identify gene relationships, networks, and epistatic interactions on a systems level. Today, high-throughput experimental tools are available to measure molecular and biochemical data. For example, DNA microarrays allow scientists to gather hundreds of thousands of data points from cells, with transcription level used as the measured phenotype. Then, computational and bioinformatics methods can be used to sift and sort though the massive amounts of biological data to search for epistatic interactions. Once we identify and understand epistatic relationships using techniques such as these, we can apply this knowledge to better diagnose and treat complex diseases.

Genome

In modern molecular biology and genetics, the genome is the entirety of an organism's hereditary information. It is encoded either in DNA or, for many types of virus, in RNA. The genome includes both the genes and the non-coding sequences of the DNA.

Origin of Term

The term was adapted in 1920 by Hans Winkler, Professor of Botany at the University of Hamburg, Germany. In Greek, the word *genome* means I become, I am born, to come into being. The Oxford English Dictionary suggests the name to be a blend of the words *gene* and *chromosome*. A few related-*ome* words already existed, such as *biome* and *rhizome*, forming a vocabulary into which *genome* fits systematically.

Overview

Some organisms have multiple copies of chromosomes, diploid, triploid, tetraploid and so on. In classical genetics, in a sexually reproducing organism (typically eukarya) the gamete has half of the number of chromosome of the somatic cell and the genome is a full set of chromosomes in a gamete. In haploid organisms, including cells of bacteria, archaea, and in organelles including mitochondria and chloroplasts, or viruses, that similarly contain genes, the single or set of circular and/or linear chains of DNA (or RNA for some viruses), likewise constitute the *genome*.

The term genome can be applied specifically to mean that stored on a complete set of *nuclear DNA* (i.e., the "nuclear genome") but can also be applied to that stored within organelles that contain their own DNA, as with the "mitochondrial genome" or the "chloroplast genome". Additionally, the genome can comprise nonchromosomal genetic elements such as viruses, plasmids, and transposable elements. When people say that the genome of a sexually reproducing species has been "sequenced", typically they are referring to a determination of the sequences of one set of autosomes and one of each type of sex chromosome, which together represent both of the possible sexes. Even in species that exist in only one sex, what is described as "a genome sequence" may be a composite read from the chromosomes of various individuals. In general use, the phrase "genetic makeup" is sometimes used conversationally to mean the genome of a particular individual or organism.

The study of the global properties of genomes of related organisms is usually referred to as genomics, which distinguishes it from genetics which generally studies the properties of single genes or groups of genes. Both the number of base pairs and the number of genes vary widely from one species to another, and there is only a rough correlation between the two (an observation known as the C-value paradox). At present, the highest known number of genes is around 60,000, for the protozoan causing trichomoniasis, almost three times as many as in the human genome.

An analogy to the human genome stored on DNA is that of instructions stored in a library:

- The library would contain 46 books (chromosomes)
- The books range in size from 400 to 3340 pages (genes)
- which is 48 to 250 million letters (A,C,G,T) per book.
- Hence the library contains over six billion letters total;
- The library fits into a cell nucleus the size of a pinpoint;
- A copy of the library (all 46 books) is contained in almost every cell of our body.

Types

Most biological entities that are more complex than a virus sometimes or always carry additional genetic material besides that which resides in their chromosomes. In some contexts, such as sequencing the genome of a pathogenic microbe, "genome" is meant to include information stored on this auxiliary material, which is carried in plasmids. In such circumstances then, "genome" describes all of the genes and information on non-coding DNA that have the potential to be present.

In eukaryotes such as plants, protozoa and animals, however, "genome" carries the typical connotation of only information on chromosomal DNA. So although these organisms contain chloroplasts and/or mitochondria that have their own DNA, the genetic information contained by DNA within these organelles is not considered part of the genome. In fact, mitochondria are sometimes said to have their own genome often referred to as the "mitochondrial genome". The DNA found within the chloroplast may be referred to as the "plastome".

Genomes and Genetic Variation

Note that a genome does not capture the genetic diversity or the genetic polymorphism of a species. For example, the human genome sequence in principle could be determined from just half the information on the DNA of one cell from one individual. To learn what variations in genetic information underlie particular traits or diseases requires comparisons across individuals. This point explains the common usage of "genome" (which parallels a common usage of "gene") to refer not to the information in any particular DNA sequence, but to a whole family of sequences that share a biological context.

Although this concept may seem counter intuitive, it is the same concept that says there is no particular shape that is the shape of a cheetah. Cheetahs vary, and so do the sequences of their genomes. Yet both the individual animals and their sequences share commonalities, so one can learn something about cheetahs and "cheetah-ness" from a single example of either.

Sequencing and Mapping

The Human Genome Project was organized to map and to sequence the human genome. Other genome projects include mouse, rice, the plant *Arabidopsis thaliana*, the puffer fish, bacteria like E. coli, etc. In 1976, Walter Fiers at the University of Ghent (Belgium) was the first to establish the complete nucleotide sequence of a viral RNA-genome (bacteriophage MS2). The first DNA-genome project to be completed was the Phage Ö-X174, with only 5386 base pairs, which was sequenced by Fred Sanger in 1977. The first bacterial genome to be completed was that of Haemophilus influenzae, completed by a team at The Institute for Genomic Research in 1995.

The development of new technologies has dramatically decreased the difficulty and cost of sequencing, and the number of complete genome sequences is rising rapidly. Among many genome database sites, the one maintained by the US National Institutes of Health is inclusive.

These new technologies open up the prospect of personal genome sequencing as an important diagnostic tool. A major step towards that goal was the May 2007 *New York Times* announcement that the full genome of DNA pioneer James D. Watson was deciphered.

Whereas a genome sequence lists the order of every DNA base in a genome, a genome map identifies the landmarks. A genome map is less detailed than a genome sequence and aids in navigating around the genome.

Genome Evolution

Genomes are more than the sum of an organism's genes and have traits that may be measured and studied without reference to the details of any particular genes and their products. Researchers compare traits such as *chromosome number* (karyotype), genome size, gene order, codon usage bias, and GC-content to determine what mechanisms could have produced the great variety of genomes that exist today. Duplications play a major role in shaping the genome. Duplications may range from extension of short tandem repeats, to duplication of a cluster of genes, and all the way to duplications of entire chromosomes or even entire genomes. Such duplications are probably fundamental to the creation of genetic novelty.

Horizontal gene transfer is invoked to explain how there is often extreme similarity between small portions of the genomes of two organisms that are otherwise very distantly related. Horizontal gene transfer seems to be common among many microbes. Also, eukaryotic cells seem to have experienced a transfer of some genetic material from their chloroplast and mitochondrial genomes to their nuclear chromosomes.

Chapter 2

Identification of Genetic Material

The genetic material of a cell or an organism refers to those materials found in the nucleus, mitochondria and cytoplasm, which play a fundamental role in determining the structure and nature of cell substances, and capable of self-propagating and variation.

When it became evident that the chromosomes were the organs of heredity, because (i) they formed the only link between two generations; (ii) they carried linearly arranged genes; (iii) they occurred in every organism in specific number and had specific morphology for a particular species; and (iv)any variation in their number or morphology affected the phenotype of the species.

Then, various attempts were made by early molecular geneticists to identify the physical and chemical nature of genes. But, the genes were found so minute structures that their physical identity remained almost impossible. However, the extensive chemical analysis of chromosomes of different organisms have revealed that chromosomes contain proteins and nucleic acid (DNA and RNA) and it was thought that genes might have either proteins or nucleic acids as their component molecules. Early molecular geneticists have assigned the informational roles of genes to the chromosomal proteins, because, they found nucleic acids too simple to carry genetic informations. The controversy about the assignment of genetic role either to chromosomal proteins or to chromosomal DNA, existed upto 1950, when finally it was universally accepted that DNA is the genetic substance (I. e., chemical of which genes are composed) of most micro-organisms and higher organisms.Later on, RNA was found to be the genetic material of some viruses.

Direct Evidences for DNA as the Genetic Material

The most conclusive evidences in support of DNA as the genetic material comes from following three avenues of approach on microorganisms: transformation of bacteria mode of infection of bacteriophages and conjugation in bacteria. Bacterial Transformation and Griffith EffectIn 1928, Frederick Griffith encountered a phenomenon now known as genetic transformation. Colonies of virulent strain (pathogenic) of pneumonia causing spherically-shaped bacterium, Diplococcus pneumoniae grown on nutrient agar, have a smooth (s) glistering appearance owing to the presence of a type specific, ploysaccharide (a polymer of glucose and glucuronic acid) capsule.

The avirulant (non-pathogenic) strains of this bacterium, on the other hand, lack this capsule and they produce dull, rough (R) colonies and is unable to cause the pneumonia disease. Smooth (S) and rough (R) characters are directly related to the presence or absence of the capsule and this trait is known to be genetically determined. Both S and R forms occur in several types and are designed as S-1, S-II, S-III, etc., and R-I, R-II, R-III, etc., respectively.

All these subtypes of S and R bacteria differ with each other in the type of antigens, they produce. The kind of antigens produced is likewise genetically determined. Smooth (S) forms sometimes mutate to rough (R) forms, but this change have not been found reversible.

In the course of his work, Griffith injected laboratory mice with live R-II pneumococci; the mice suffered no illness because R-II pneumococci was avirulent.

But when the mice were injected with the mixture of living avirulent R-II and heat killed S-III virulent, the unexpected symptoms of pneumonia appeared and high mortality was resulted in them.

By postmortoming the dead mice, it was found that their heart blood had both R-II and S-III pneumococci. From these results, Griffith concluded that the presence of the heat-killed S-IU bacteria must have caused a transformation of the living R-II bacteria, so as to restore to them the capacity for capsule formation they had earlier lost by gene mutation. This was called "Griffith effect" or more popularly as "bacterial transformation".

Identification of the "transforming substance"-Griffith could not understand the cause of bacterial transformation and that is first of all identified by O. T. Avery, C. M. McLeod and M McCarthy (1944). They tested a fraction of heat killed S-III bacteria for the transforming ability.

They could remove proteins, lipids, polysaccharides and ribonucleic acids from the S-III extract by a variety of chemical and enzymatic methods without seriously diminishing its power to transfer R-II mutants into the S-III wild types. They ultimately concluded that because a cell free and highly purified DNA extract of S-III bacteria could bring about transformation of R-I1 bacteria into S-III, therefore, DNA is the genetic material of pneumococci. Later on, such a transforming substance, the DNA, was found in a variety of bacteria (i.e., Hemophilus influenzae. Bacillus subtilus, Shigelia paradysenteriae, etc.) and several other organisms.

Mode of Infection of Bacteriophages and Identification of DNA as their Genetic Material

By using radioactive tracers, A. Hershey and M. Chase (1952) provided further direct proof that DNA is the genetic material in certain bacterial viruses. These investigators were studying the bacteriophages that attack the bacterium, Escherichia coli. They prepared a chemically defined culture medium (for E. coli) containing phosphoric acid and sulphuric acid as the only source of phosphorus and sulphur.

Known quantities of radioactive isotopes of phosphorus (P^{32}) and sulphur (S^{32}) were added to the culture medium. E. coli grown on such a culture medium incorporated the P^{32} and S^{35} into their chemical constituents. When T_2 bacteriophage particles infected the radioactive bacteria, they incorporated the labeled DNA.

Phage capsid proteins did not contain appreciable amounts of phosphorus and only the DNA was labeled with P^{32}. Similarly, the protein envelope around the phage was selectively labeled with S^{35}. DNA contains virtually no sulphur and was not labeled with S^{35} By this method it was possible to differentially label the phage DNA and proteins of phage-capsid.

After phage growth, the phage particles were separated from the host cell by centrifugation. The radioactive phages were next introduced to non-radioactive bacterial cultures where they attacked the bacteria. Subsequently the viruses were separated from the host cell by agitation and the content of P^{32} and S^{35} in the host (E. coli) and parasite (bacteriophage) was determined. The phosphorus label was found to be associated with the bacterial cells and the sulphur label was in the proteins coats (capsids) left in the medium. This indicated that the DNA had panetrated the cells but that the protein coat or capsid of the phage was left outside the wall of the bacterium. The labeled DNA was found to reproduced in the host cells. These experiments, thus, clearly demonstrated that genetic informations of these bacteriophages resided in their DNA molecules and not in the protein molecules of their capsids.

Later on Hershey-Chase experiment was repeated for other viruses and it was found that DNA is also the genetic material in a number of other viruses.

Bacterial Conjugation

Another conclusive evidence for DNA as the genetic material came from the phenomenon of conjugation of bacteria. Laderberg and Tatum (1946) found that when a F$^+$ ('male') E. coli cell conjugated with a F$^-$ ('female') E. coli cell an unidirectional transfer of F$^+$ factor of 'male' cell to F$^-$ or 'female' cell took place; so that the latter was converted into a F$^+$ or 'male' strain. The F$^+$ factor was found to be a fragment of DNA molecule which occurred in the cytoplasm of bactarial cell.

Indirect Evidences for DNA as the Genetic Material

The indirect or circumstantial evidences in favour of DNA as the genetic material came from higher organisms which are not as easy to manipulate as bacteria and viruses. The fact that DNA is the genetic material of higher organisms has been supported by following facts:

1. Localization -All cells contain DNA which is localized primarily on chromosomes in the nucleus of the cell. This is consistent with the fact the genes are also located specifically on chromosomes. The Feulgen techniques have shown that DNA entirely remains restricted to the chromosomes and it forms one of the major component of chromosomes.

2. Quantity-Various quantitative measurement of amount of DNA in different cells have shown that there is a correlation between the amount of DNA and the number of chromosome sets (ploidy).

Cells set	Mean DNA Fegulen content	Presumed chromosomes
	(Picograms)	(Ploidy)
Spermatids	1.68	Haploid(n)
Liver	3.16	Diploid(2n)
Liver	6.30	Tetraploid(4n)
Liver	12.80	Octoploid(8n)

S. No	Organism	Kidney	Liver	Erythrocytes	Sperm
1.	Chicken	2.4	2.5	2.5	1.3
2.	Bovine	6.4	6.4	-	3.3
3.	Carp	-	3.0	3.3	1.6
4.	Human	5.4	5.6	2.5	2.5

The diploid amount of DNA is constant within the species but varies from one species to another, as shown in.

Stability-Most of the macromolecules of the cell are constantly being broken down and synthesized. If this were to happen to a gene, valuable hereditary information would invariably be lost. Of all the macromolecules in the cell, DNA is the most metabolically stable.

Sensitivity to chemical and physical agents-All agents which cause genes to be damaged (mutation) have also been shown to induce changes in the structure of DNA. For example, one such mutagenic agent, ultraviolet light, is capable of producing breaks in the DNA molecule.

In 1950 Swift found that the amount of DNA per resting nucleus (interphase nucleus) in mitotically active tissues of animals ranged from twice (2n) the value found in the gametes (In) to four times (4n) this value. He concluded that DNA synthesis occurred during the interphase.

This parallelism of behaviour in DNA and chromosomes clearly indicates that DNA is the genetic material of higher animals.

Evidences for RNA as the Genetic Material of Some Viruses

The demonstration that RNA is the genetic material in RNA containing viruses came in 1956, when A. Gierer and G. Schramm showed that tobacco plants could be inoculated with purified RNA from the tobacco mosaic virus (TMV) and TMV -like lesions could later be identified on the tobacco leaves.

A different approach was taken by H. Fraenkel Conrat and B. Singer in experiments published in 1957. They first separated the RNA from the protein of TMV viruses, much as Hershey and Chase had done with the DNA and the protein of T2 bacteriophage. They then developed techniques for forming 'reconstituted' viruses containing the protein from one mutant strain of TMV and the RNA from another or vice-versa.

Such hybrid viruses were allowed to infect tobacco leaves, and the progeny were examined. In all cases the progeny were the parental RNA type and not the parental protein type.

In a much improved technique N. Pace and S. Spiegelman in 1966 purified RNA from two different mutant strains of the RNA phage QB which had quite distinct base compositions.

The isolated RNA'S were then incubated separately in the presence of an E. coli cell extract containing an enzyme capable of RNA replication. The new RNA synthesized was in each case identical in base composition to particular phage RNA presented to the in vitro system, thus indicating that the phage RNA-can .serve as a template for its self-replication.

Base Pair

In molecular biology and genetics, two nucleotides on opposite complementary DNA or RNA strands that are connected via hydrogen bonds are called a base pair (often abbreviated bp). In the canonical Watson-Crick DNA base pairing, adenine (A) forms a base pair with thymine (T) and guanine (G) forms a base pair with cytosine (C). In RNA, thymine is replaced by uracil (U).

Alternate hydrogen bonding patterns, such as the wobble base pair and Hoogsteen base pair, also occur—in particular, in RNA—giving rise to complex and functional tertiary structures. It is important to note that pairing is the mechanism by which codons on messenger RNA molecules are recognized by anticodons on transfer RNA during protein translation. Some DNA- or RNA-binding enzymes can recognize specific base pairing patterns that identify particular regulatory regions of genes.

The size of an individual gene or an organism's entire genome is often measured in base pairs because DNA is usually double-stranded. Hence, the number of total base pairs is equal to the number of nucleotides in one of the strands (with the exception of non-coding single-stranded regions of telomeres).

The haploid human genome (23 chromosomes) is estimated to be about 3 billion base pairs long and to contain 20,000–25,000 distinct genes. A kilobase (kb) is a unit of measurement in molecular biology equal to 1000 base pairs of DNA or RNA.

Hydrogen Bonding and Stability

Hydrogen bonding is the chemical interaction that underlies the base-pairing rules described above. Appropriate geometrical correspondence of hydrogen bond donors and acceptors allows only the "right" pairs to form stably. DNA with high GC-content is more stable than DNA with low GC-content, but, contrary to popular belief, the hydrogen bonds do not stabilize the DNA significantly, and stabilization is mainly due to stacking interactions.

The larger nucleobases, adenine and guanine, are members of a class of double-ringed chemical structures called purines; the smaller nucleobases, cytosine and thymine (and uracil), are members of a class of single-ringed chemical structures called pyrimidines. Purines are complementary only with pyrimidines: pyrimidine-pyrimidine pairings are energetically unfavourable because the molecules are too far apart for hydrogen bonding to be established; purine-purine pairings are energetically unfavourable because the molecules are too close, leading to overlap repulsion. The only other possible pairings are GT and AC; these pairings are mismatches because the pattern of hydrogen donors and acceptors do not correspond. The GU pairing, with two hydrogen bonds, does occur fairly often in RNA.

Paired DNA and RNA molecules are comparatively stable at room temperature but the two nucleotide strands will separate above a melting point that is determined by the length of the molecules, the extent of mispairing (if any), and the GC content. Higher GC content results in higher melting temperatures; it is, therefore, unsurprising that the genomes of extremophile organisms such as *Thermus thermophilus* are particularly GC-rich. On the converse, regions of a genome that need to separate frequently — for example, the promoter regions for often-transcribed genes — are comparatively GC-poor. GC content and melting temperature must also be taken into account when designing primers for PCR reactions.

Base Stacking

Base stacking interactions in DNA and RNA are due to dispersion attraction, short-range exchange repulsion, and electrostatic interactions, which also contribute to stability. Again, GC stacking interactions with adjacent bases tend to be more favourable. (Note, however, that a GC stacking interaction with the next base pair is geometrically different from a CG interaction.) Base stacking effects are especially important in the secondary structure and tertiary structure of RNA; for example, RNA stem-loop structures are stabilized by base stacking in the loop region.

Nucleic Acid Analogues

Nucleic acid analogues are compounds structurally similar (analog) to naturally occurring RNA and DNA, used in medicine and in molecular biology research. Nucleic acids are chains of nucleotides, which are composed of three parts: a phosphate backbone, a pucker-shaped pentose sugar, either ribose or deoxyribose, and one of four nucleobases. An analogue may have any of these altered. Typically the analogue nucleobases confer, among other things, different base pairing and base stacking proprieties. Examples include universal bases, which can pair with all four canon bases, and phosphate-sugar backbone analogues such as PNA, which affect the properties of the chain (PNA can even form a triple helix).

Artificial nucleic acids include peptide nucleic acid (PNA), Morpholino and locked nucleic acid (LNA), as well as glycol nucleic acid (GNA) and threose nucleic acid (TNA). Each of these is distinguished from naturally-occurring DNA or RNA by changes to the backbone of the molecule.

Medicine

Several nucleoside analogues are used as antiviral or anticancer agents. The viral polymerase incorporates these compounds with non-canon bases. These compounds are activated in the cells by being converted into nucleotides, they are administered as nucleosides since charged nucleotides cannot easily cross cell membranes.

Molecular Biology

Nucleic acid analogues are used in molecular biology for several purposes:
- As a tool to detect particular sequences
- As a tool with resistance to RNA hydrolysis
- As a tool for another purpose, such as sequencing
- Naturally occurring, such as in tRNA
- Investigation of the mechanisms used by enzyme, such as an Enzyme inhibitor
- Investigation of possible scenarios of the origin of life

- Investigation of the structural features of nucleic acids
- Investigation of the possible alternatives to the natural system in Synthetic biology.

Backbone Analogues

Hydrolysis Resistant RNA-analogues: To overcome the fact that ribose's 2' hydroxy group that reacts with the phosphate linked 3' hydroxy group (RNA is too unstable to be used or synthesized reliably), a ribose analogue is used. The most common RNA analogues are locked nucleic acid (LNA), morpholino, and peptide nucleic acid (PNA). These oligonucleotides differ as they have a different backbone sugar but still bind according to Watson and Crick pairing with RNA or DNA, but are immune to nuclease activity (They generally cannot be enzymatically synthesized and can only be produced synthetically).

Other Notable Analogues Used as Tools

Dideoxynucleotides are used in sequencing. These nucleoside triphosphates possess a non-canonical sugar, dideoxyribose, which lacks the 3' hydroxyl group normally present in DNA and therefore cannot bond with the next base. The lack of the 3' hydroxyl group terminates the chain reaction as the DNA polymerases mistake it for a regular deoxyribonucleotide. Another chain-terminating analogue that lacks a 3' hydroxyl and mimics adenosine is called cordycepin. Cordycepin is an anticancer drug that targets RNA replication. Another analogue in sequencing is a nucleobase analogue, 7-deaza-GTP and is used to sequence CG rich regions, instead 7-deaza-ATP is called tubercidin, an antibiotic.

Precursors to the RNA World

RNA may be too complex to be the first nucleic acid, so before the RNA world several simpler nucleic acids that differ in the backbone, such as TNA and GNA and PNA, have been offered as candidates for the first nucleic acids.

Base Analogues

Nucleobase Structure and Nomenclature: Natural bases are divided into two classes depending on their structure: pyrimidine (an heterocyclic aromatic six-membered ring with nitrogen atoms in position 1 and 3) and purine (a pyrimidine (numeration inverted) fused with an imidazole ring, a five-membered ring with 2 nitrogen atoms separated by one carbon (meta), 7,9). Their main proprieties are base pairing, resulting form 2 or 3 hydrogen bonds between ketone (electron withdrawing group,*ei.* more negatively charged) and amino (electron withdrawing group,*ei.* more positively charged) functional groups, and base stacking, caused by the attraction of the delocalized ∂ electron clouds of the aromatic ring structure.

Fluorophores

Commonly fluorophores (such as rhodamine or fluorescein) are linked to the ring linked to the sugar (in para) via a flexible arm, presumably extruding from the major groove of the helix. Due to low processivity of the nucleotides linked to bulky adducts such as florophores by taq polymerases, the sequence is typically copied using a nucleotide with an arm and later coupled with a reactive fluorophore (indirect labelling):

- amine reactive: Aminoallyl nucleotide contain a primary amine group on a linker that reacts with the amino-reactive dye such as a cyanine or Alexa Fluor dyes, which contain a reactive leaving group, such as a succinimidyl ester (NHS). (base-pairing amino groups are not affected).

- thiol reactive: thiol containing nucleotides reacts with the fluorophore linked to a reactive leaving group, such as a maleimide.

- biotin linked nucleotides rely on the same indirect labelling principle (+ fluorescent streptavidin) and are used in Affymetrix DNAchips.

Fluorophores find a variety of uses in medicine and biochemistry.

Fluorescent Base Analogues

The most commonly used and commercially available fluorescent base analogue, 2-aminopurine (2-AP), has a high-fluorescence quantum yield free in solution (0.68) that is considerably reduced (appr. 100 times but highly dependent on base sequence) when incorporated into nucleic acids. The emission sensitivity of 2-AP to immediate surroundings is shared by other promising and useful fluorescent base analogues like 3-MI, 6-MI, 6-MAP, pyrrolo-dC (also commercially available), modified and improved derivatives of pyrrolo-dC, furan-modified bases and many other ones. This sensitivity to the microenvironment have been utilized in studies of e.g. structure and dynamics within both DNA and RNA, dynamics and kinetics of DNA-protein interaction and electron transfer within DNA. A newly developed and very interesting group of fluorescent base analogues that has a fluorescence quantum yield that is nearly insensitive to their immediate surroundings is the tricyclic cytosine family. 1,3-Diaza-2-oxophenothiazine, tC, has a fluorescence quantum yield of approximately 0.2 both in single- and in double-strands irrespective of surrounding bases. Also the oxo-homologue of tC called tC^O (both commercially available), 1,3-diaza-2-oxophenoxazine, has a quantum yield of 0.2 in double-stranded systems. However, it is somewhat sensitive to surrounding bases in single-strands (quantum yields of 0.14–0.41). The high and stable quantum yields of these base analogues make them very bright, and, in combination with their good base analogue properties (leaves DNA structure and stability next to unperturbed), they are especially useful in fluorescence anisotropy and FRET measurements, areas where other fluorescent base analogues are less accurate. Also, in the same family of cytosine analogues, a FRET-acceptor base analogue, tC$_{nitro}$, has been developed. Together with tC^O as a FRET-donor this constitutes the first nucleic acid base analogue FRET-pair ever developed. The tC-family has, for example, been used in studies related to polymerase DNA-binding and DNA-polymerization mechanisms.

Natural Non-canon Bases

In a cell, there are several noncanon bases present: CpG islands in DNA (are often methylated), all eukaryotic mRNA (capped with a methyl-7-guanosine), and several bases of rRNAs (are methylated). Often, tRNAs are heavily modified posttranscriptionally in order to improve their conformation or base pairing, in particular in/near the anticodon: inosine can base pair with C, U, and even with A, whereas thiouridine (with A) is more specific than uracil (with a purine). Other common tRNA base modifications are pseudouridine (which gives its name to the TØC loop), dihydrouridine (which does not stack as it is not aromatic), queuosine, wyosine, and so forth. Nevertheless these are all modifications to normal bases and are not placed by a polymerase.

Base-pairing

Canonical bases may have either a ketone or an amine group on the carbons surrounding the nitrogen atom furthest away from the glycosidic bond, which allows them to base pair (Watson-Crick base pairing) via hydrogen bonds (amine with ketone, purine with pyrimidine). Adenine and 2-aminoadenine have one/two amine group(s), whereas thymine has two ketone groups, and cytosine and guanine are mixed amine and ketone (inverted in respect to each other). The precise reason why there are only four nucleotides is debated, but there are several unused possibilities.

Furthermore, adenine is not the most stable choice for base pairing: in Cyanophage S-2L diaminopurine (DAP) is used instead of adenine (host evasion). Diaminopurine basepairs perfectly with thymine as it is identical to adenine but has an amine group at position 2 forming 3 intramolecular hydrogen bonds, eliminating the major difference between the two types of basepairs (Weak:A-T and Strong:C-G). This improved stability affects protein-binding ineractions that rely on those differences. Other combination include,

- isoguanine and isocytosine, which have their amine and ketone inverted compared to standard guanine and cytosine, (not used probably as tautomers are problematic for base pairing, but isoC and isoG can be amplified correctly with PCR even in the presence of the 4 canon bases)
- diaminopyrimidine and a xanthine, which bind like 2-aminoadenine and thymine but with inverted structures (not used as xanthine is a deamination product).

However, correct DNA structure can form even when the bases are not paired via hydrogen bonding; that is, the bases pair thanks to hydrophobicity, as studies have shown using DNA isosteres (analogues with same number of atoms), such as the thymine analogue 2,4-difluorotoluene (F) or the adenine analogue 4-methylbenzimidazole (Z). An alternative hydrophobic pair could be isoquinoline, and the pyrrolo[2,3-b]pyridine.

Other noteworthy basepairs:

- Several fluorescent bases have also been made, such as the 2-amino-6-(2-thienyl)purine and pyrrole-2-carbaldehyde base pair.
- Metal coordinated bases, such as two 2,6-bis (ethylthiomethyl) pyridine (SPy) with a silver ion or pyridine-2,6-dicarboxamide (Dipam) and a mondentate pyridine (Py) with a copper ion.
- Universal bases may pair indiscriminately with any other base, but, in general, lower the melting temperature of the sequence considerably; examples include 2'-deoxyinosine (hypoxanthine deoxynucleotide) derivatives, nitroazole analogues, and hydrophobic aromatic non-hydrogen-bonding

bases (strong stacking effects). These are used as proof of concept and, in general, are not utilised in degenerate primers (which are a mixture of primers).

- The numbers of possible base pairs is doubled when xDNA is considered. xDNA contains expanded bases, in which a benzene ring has been added, which may pair with canon bases, resulting in four possible base-pairs (8 bases:xA-T,xT-A,xC-G,xG-C, 16 bases if the unused arrangements are used). Another form of benzene added bases is yDNA, in which the base is widened by the benzene.

Metal Base Pairs

In metal base-pairing, the Watson-Crick hydrogen bonds are replaced by the interaction between a metal ion with nucleosides acting as ligands. The possible geometries of the metal that would allow for duplex formation with two bidentate nucleosides around a central metal atom are: tetrahedral, dodecahedral, and square planar. Metal-complexing with DNA can occur by the formation of non-canonical base pairs from natural nucleobases with participation by metal ions and also by the exchanging the hydrogen atoms that are part of the Watson-Crick base pairing by metal ions. Introduction of metal ions into a DNA duplex has shown to have potential magnetic, conducting properties, as well as increased stability.

Metal complexing has been shown to occur between natural nucleobases. A well-documented example is the formation of T-Hg-T, which involves two deportonated thymine nucleobases that are brought together by Hg^{2+} and forms a connected metal-base pair. This motif does not accommodate stacked Hg^{2+} in a duplex due to an intrastrand hairpin formation process that is favoured over duplex formation.

Two thymines across from each other in a duplex do not form a Watson-Crick base pair in a duplex; this is an example where a Watson-Crick basepair mismatch is stabilized by the formation of the metal-base pair. Another example of a metal complexing to natural nucleobases is the formation of A-Zn-T and G-Zn-C at high pH; Co^{+2} and Ni^{+2} also form these complexes. These are Watson-Crick base pairs where the divalent cation in coordinated to the nucleobases. The exact binding is debated.

A large variety of artificial nucleobases have been developed for use as metal base pairs. These modified nucleobases exhibit tunable electronic properties, sizes, and binding affinities that can be optimized for a specific metal. For, example a nucleoside modified with a pyridine-2,6-dicarboxylate has shown to bind tightly to Cu^{2+}, whereas other divalent ions are only loosely bound. The tridentate character contributes to this selectivity.

The fourth coordination site on the copper is saturated by an oppositely arranged pyridine nucleobase. The asymmetric metal base pairing system is orthogonal to the Watson-Crick base pairs. Another example of an artificial nucleobase is that with hydroxypyridone nucleobases, which are able to bind Cu^{2+} inside the DNA duplex.

Five consecutive copper-hydroxypyridone base pairs were incorporated into a double strand, which were flanked by only one natural nucleobase on both ends. EPR data showed that the distance between copper centres was estimated to be 3.7 ± 0.1 Å, while a natural B-type DNA duplex is only slightly larger (3.4 Å). The appeal for stacking metal ions inside a DNA duplex is the hope to obtain nanoscopic self-assembling metal wires, though this has not been realized yet.

Orthogonal System

It has been proposed and studied both theoretically and experimentally the possibility of implementing an orthogonal system inside cells independent of the cellular genetic material in order to make a completely safe system, with the possible increase in encoding potentials Several groups have focused on different aspects:

- novel backbones and base pairs as discussed above
- XNA replication/transcription polymerases starting generally from T7 RNA polymerase
- rybozomes (16S sequences with altered anti Shine-Dalgarno sequence allowing the translation of only orthogonal mRNA with a matching altered Shine-Dalgarno sequence)
- novel tRNA encoding non-natural aminoacids.

Base Analogs and Intercalators

Chemical analogs of nucleotides can take the place of proper nucleotides and establish non-canonical base-pairing, leading to errors (mostly point mutations) in DNA replication and DNA transcription. This is due to their isosteric chemistry. One common mutagenic base analog is 5-bromouracil, which resembles thymine but can base-pair to guanine in its enol form.

Other chemicals, known as DNA intercalators, fit into the gap between adjacent bases on a single strand and induce frameshift mutations by "masquerading" as a base, causing the DNA replication machinery to skip or insert additional nucleotides at the intercalated site. Most intercalators are large polyaromatic compounds and are known or suspected carcinogens. Examples include ethidium bromide and acridine.

Examples

The following DNA sequences illustrate pair double-stranded patterns. By convention, the top strand is written from the 5' end to the 3' end; thus, the bottom strand is written 3' to 5'.

A base-paired DNA sequence:

ATCGATTGAGCTCTAGCG

TAGCTAACTCGAGATCGC

The corresponding RNA sequence, in which uracil is substituted for thymine where uracil takes its place in the RNA strand:

AUCGAUUGAGCUCUAGCG

UAGCUAACUCGAGAUCGC

Length Measurements

The following abbreviations are commonly used to describe the length of a D/RNA molecule:

- bp = base pair(s)—one bp corresponds to circa 3.4 Å of length along the strand
- kb (= kbp) = kilo base pairs = 1,000 bp
- Mb = mega base pairs = 1,000,000 bp
- Gb = giga base pairs = 1,000,000,000 bp.

In case of single stranded DNA/RNA units of nucleotides are used, abbreviated nt (or knt, Mnt, Gnt), as they are not paired. For distinction between units of computer storage and bases kbp, Mbp, Gbp, etc. may be used for basepairs. The length of 16S rDNA for bacteria is 1542 base-pairs in length. The Centimorgan is also often used to imply distance along a chromosome, but the number of base-pairs it corresponds to varies widely. In the Human genome, it is about 1 million base pairs.

Protein Synthesis

Process whereby DNA encodes for the production of amino acids and proteins.

Transcription

Before the synthesis of a protein begins, the corresponding RNA molecule is produced by RNA transcription. One strand of the DNA double helix is used as a template by the RNA polymerase to synthesize a messenger RNA (mRNA). This mRNA migrates from the nucleus to the cytoplasm. During this step, mRNA goes through different types of maturation including one called splicing when the non-coding sequences are eliminated. The coding mRNA sequence can be described as a unit of three nucleotides called a codon.

Translation

The ribosome binds to the mRNA at the start codon (AUG) that is recognized only by the initiator tRNA. The ribosome proceeds to the elongation phase of protein synthesis. During this stage, complexes, composed of an amino acid linked to tRNA, sequentially bind to the appropriate codon in mRNA by forming complementary base pairs with the tRNA anticodon. The ribosome moves from codon to codon along the mRNA. Amino acids are added one by one, translated into polypeptidic sequences dictated by DNA and represented by mRNA. At the end, a release factor binds to the stop codon, terminating translation and releasing the complete polypeptide from the ribosome. One specific amino acid can correspond to more than one codon. The genetic code is said to be degenerate.

DNA and Protein Synthesis

With over 100,000 different proteins to manufacture, how the heck does our body get it right? When one thinks of the amount of information the body needs to keep track of, - eye, hair and skin colour, protein sequence, toenail size, etc. - it would seem a task for a supercomputer to record all of the necessary information. In essence it is. But not a supercomputer made of silicon wafers and TV screens, rather one made of an intricate biomolecule called DNA.

DNA (deoxyribonucleic acid) is in the family of molecules referred to as nucleic acids. One strand of DNA has a backbone consisting of a polymer of the simple sugar deoxyribose bonded to something called a phosphate unit.

What is impressive about DNA is that each sugar molecule in the strand also binds to one of four different nucleotide bases. These bases: Adenine (A), Guanine (G), Cytosine (C) and Thymine (T), are the beginnings of what we will soon see is a molecular alphabet. Each sugar molecule in the DNA strand will bind to one nucleotide base. Thus, as our description of DNA unfolds, we see that a single strand of the molecule looks more like this:

C		T		G		A	
sugar-	phosphate-	sugar-	phosphate-	sugar-	phosphate-	sugar-	phosph

Each strand of DNA contains millions or even billions (in the case of human DNA) of nucleotide bases. These bases are arranged in a specific order according to our genetic ancestry. The order of these base units makes up the code for specific characteristics in the body, such as eye colour or nose-hair length. Just as we use 26 letters in various sequences to code for the words you are now reading, our body's DNA uses 4 letters (the 4 nucleotide bases) to code for millions of different characteristics.

Each molecule of DNA is actually made up of 2 strands of DNA cross-linked together. Each nucleotide base in the DNA strand will cross-link (via hydrogen bonds) with a nucleotide base in a second strand of DNA forming a structure that resembles a ladder. These bases cross-link in a very specific order: A will only link with T (and vice-versa), and C will only link with G (and vice-versa). Thus our picture of DNA now looks like this:

sugar-	phosphate-	sugar-	phosphate-	sugar-	phosphate-	sugar-	phosph
G		A		C		T	
C		T		G		A	
sugar-	phosphate-	sugar-	phosphate-	sugar-	phosphate-	sugar-	phosph

In 1953, James Watson, Francis Crick and Rosalind Franklin discovered that the structure of DNA is actually a double helix. In other words, the DNA ladder described above coils around itself somewhat like the cord of a telephone, as illustrated at right. To get a better picture of the DNA double helix, Dr. Abby Parrill and Dr. Jacquelyn Gervay at the University of Arizona have put together a movie of the DNA double helix and an interactive strand of DNA that can be moved around by the user. Both are available by clicking below (Note: the interactive DNA strand requires Chemscape's CHIME software to run).

Bottom of Form

The specific base-pairing of DNA aids in reproduction of the double helix when more genetic material is needed (such as during reproduction, to pass on characteristics from parent to offspring).

When DNA reproduces, the 2 strands unzip from each other and enzymes add new bases to each, thus forming two new strands. This process is illustrated in the Access Excellence DNA Replicating Itself page (just hit your browser's Back button to return here).

Within this coil of DNA lies all the information needed to produce everything in the human body. A strand of DNA may be millions, or billions, of base-pairs long. Different segments of the DNA molecule code for different characteristics in the body. A Gene is a relatively small segment of DNA that codes for the synthesis of a specific protein. This protein then will play a structural or functional role in the body. A chromosome is a larger collection of DNA that contains many genes and the support proteins needed to control these genes.

Protein Synthesis

How does a gene code for a protein? Protein synthesis is a 2 part process that involves a second type of nucleic acid along with DNA. This second type of nucleic acid is RNA, ribonucleic acid. RNA differs from DNA in two respects. First, the sugar units in RNA are ribose as compared to DNA's deoxyribose. Because of this difference, RNA does not bind to the nucleotide base Thymine, instead, RNA contains the nucleotide base Uracil (U) in place of T (RNA also contains the other three bases: A, C and G).

Transcription: In the first step of protein synthesis, the 2 DNA strands in a gene that codes for a protein unzip from each other. Similar to the way DNA replicates itself, a single strand of messenger RNA (mRNA) is then made by pairing up mRNA bases with the exposed DNA nucleotide bases.

Remember that mRNA does not contain the base Thymine, so U is paired with each of DNA's A bases.

Translation: After the mRNA is manufactured, it leaves the cell nucleus and travels to a cellular organelle called the ribosome (we will learn about the cell, nucleus and ribosome in the next lesson). In the ribosome, the mRNA code is translated into a transfer RNA (tRNA) code which, in turn, is transfered into a protein sequence. In this process, each set of 3 mRNA bases (the mRNA base triplet is called a codon) will pair with a complimentary tRNA base triplet (called an anticodon). Each tRNA is specific to an amino acid, as tRNA's are added to the sequence, amino acids are linked together by peptide bonds, eventually forming a protein that is later released by the tRNA. Using the mRNA strand we obtained above, you can generate the complimentary tRNA/amino acid sequence by clicking on the mRNA codons in the table below.

mRNA codon ⟶

After the processes of transcription and translation are complete, we are left with a protein that consists of the chain:

Although our 'protein' is only 2 amino acids in length, proteins normally consist of hundreds or thousands of amino acids.

Regulation of Gene Expression or Regulation of Protein Synthesis

Regulation of gene expression (or gene regulation) includes the processes that cells and viruses use to regulate the way that the information in genes is turned into gene products. Although a functional gene product may be an RNA or a protein, the majority of known mechanisms regulate protein coding genes. Any step of the gene's expression may be modulated, from DNA-RNA transcription to the post-translational modification of a protein.

Gene regulation is essential for viruses, prokaryotes and eukaryotes as it increases the versatility and adaptability of an organism by allowing the cell to express protein when needed. The first discovered example of a gene regulation system was the lac operon, discovered by Jacques Monod, in which protein involved in lactose metabolism are expressed by *E. coli* only in the presence of lactose and absence of glucose.

Furthermore, gene regulation drives the processes of cellular differentiation and morphogenesis, leading to the creation of different cell types in multicellular organisms where the different types of cells may possess different gene expression profiles though they all possess the same genome sequence.

Regulated Stages of Gene Expression

Any step of gene expression may be modulated, from the DNA-RNA transcription step to post-translational modification of a protein. The following is a list of stages where gene expression is regulated, the most extensively utilised point is Transcription Initiation:

- Chromatin domains.

Epigenetics

In biology, and specifically genetics, epigenetics is the study of heritable changes in phenotype (appearance) or gene expression caused by mechanisms other than changes in the underlying DNA sequence, hence the name *epi-genetics*. These changes may remain through cell divisions for the remainder of the cell's life and may also last for multiple generations. However, there is no change in the underlying DNA sequence of the organism; instead, non-genetic factors cause the organism's genes to behave (or "express themselves") differently. One example of epigenetic changes in eukaryotic biology is the process of cellular differentiation. During morphogenesis, totipotent stem cells become the various pluripotent cell lines of the embryo which in turn become fully differentiated cells. In other words, a single fertilized egg cell – the zygote – changes into the many cell types including neurons, muscle cells, epithelium, blood vessels etc. as it continues to divide. It does so by activating some genes while inhibiting others.

Etymology and Definitions

Epigenetics (as in "epigenetic landscape") was coined by C. H. Waddington in 1942 as a portmanteau of the words *genetics* and *epigenesis*. *Epigenesis* is an old word which has more recently been used to describe the differentiation of cells from their initial totipotent state in embryonic development. When Waddington coined the term the physical nature of genes and their role in heredity was not known; he used it as a conceptual model of how genes might interact with their surroundings to produce a phenotype.

Robin Holliday defined epigenetics as "the study of the mechanisms of temporal and spatial control of gene activity during the development of complex organisms." Thus *epigenetic* can be used to describe anything other than DNA sequence that influences the development of an organism.

The modern usage of the word in scientific discourse is more narrow, referring to heritable traits (over rounds of cell division and sometimes transgenerationally) that do not involve changes to the underlying DNA sequence. The Greek prefix *epi-* in *epigenetics* implies features that are "on top of" or "in addition to" genetics; thus *epigenetic* traits exist on top of or in addition to the traditional molecular basis for inheritance.

The similarity of the word to "genetics" has generated many parallel usages. The "epigenome" is a parallel to the word "genome", and refers to the overall epigenetic state of a cell. The phrase "genetic code" has also been adapted—the "epigenetic code" has been used to describe the set of epigenetic features that create different phenotypes in different cells. Taken to its extreme, the "epigenetic code" could represent the total state of the cell, with the position of each molecule accounted for in an *epigenomic map*, a diagrammatic representation of the gene expression, DNA methylation and histone modification status of a particular genomic region. More typically, the term is used in reference to systematic efforts to measure specific, relevant forms of epigenetic information such as the histone code or DNA methylation patterns.

The psychologist Erik Erikson used the term *epigenetic* in his theory of psychosocial development. That usage, however, is of primarily historical interest.

Molecular Basis of Epigenetics

The molecular basis of epigenetics is complex. It involves modifications of the activation of certain genes, but not the basic structure of DNA. Additionally, the chromatin proteins associated with DNA may be activated or silenced. This accounts for why the differentiated cells in a multi-cellular organism express only the genes that are necessary for their own activity. Epigenetic changes are preserved when cells divide. Most epigenetic changes only occur within the course of one individual organism's lifetime, but, if a mutation in the DNA has been caused in sperm or egg cell that results in fertilization, then some epigenetic changes are inherited from one generation to the next. This raises the question of whether or not epigenetic changes in an organism can alter the basic structure of its DNA, a form of Lamarckism.

Specific epigenetic processes include paramutation, bookmarking, imprinting, gene silencing, X chromosome inactivation, position effect, reprogramming, transvection, maternal effects, the progress of carcinogenesis, many effects of teratogens, regulation of histone modifications and heterochromatin, and technical limitations affecting parthenogenesis and cloning.

Epigenetic research uses a wide range of molecular biologic techniques to further our understanding of epigenetic phenomena, including chromatin immunoprecipitation (together with its large-scale variants ChIP-on-chip and ChIP-seq), fluorescent in situ hybridization, methylation-sensitive restriction enzymes, DNA adenine methyltransferase identification (DamID) and bisulphite sequencing. Furthermore, the use of bioinformatic methods is playing an increasing role (computational epigenetics).

Mechanisms

Several types of epigenetic inheritance systems may play a role in what has become known as cell memory:

DNA Methylation and Chromatin Remodeling

Because the phenotype of a cell or individual is affected by which of its genes are transcribed, heritable transcription states can give rise to epigenetic effects. There are several layers of regulation of gene expression. One way that genes are regulated is through the remodeling of chromatin. Chromatin is the complex of DNA and the histone proteins with which it associates. Histone proteins are little spheres that DNA wraps around. If the way that DNA is wrapped around the histones changes, gene expression can change as well. Chromatin remodeling is accomplished through two main mechanisms:

1. The first way is post translational modification of the amino acids that make up histone proteins. Histone proteins are made up of long chains of amino acids. If the amino acids that are in the chain are changed, the shape of the histone sphere might be modified. DNA is not completely unwound during replication. It is possible, then, that the modified histones may be carried into each new copy of the DNA. Once there, these histones may act as templates, initiating the surrounding new histones to be shaped in the new manner. By altering the shape of the histones around it, these modified histones would ensure that a differentiated cell would stay differentiated, and not convert back into being a stem cell.

2. The second way is the addition of methyl groups to the DNA, mostly at CpG sites, to convert cytosine to 5-methylcytosine. 5-Methylcytosine performs much like a regular cytosine, pairing up with a guanine. However, some areas of genome are methylated more heavily than others and highly methylated areas tend to be less transcriptionally active, through a mechanism not fully understood. Methylation of cytosines can also persist from the germ line of one of the parents into the zygote, marking the chromosome as being inherited from this parent (genetic imprinting).

The way that the cells stay differentiated in the case of DNA methylation is clearer to us than it is in the case of histone shape. Basically, certain enzymes (such as DNMT1) have a higher affinity for the methylated cytosine. If this enzyme reaches a "hemimethylated" portion of DNA (where methylcytosine is in only one of the two DNA strands) the enzyme will methylate the other half.

Although histone modifications occur throughout the entire sequence, the unstructured N-termini of histones (called histone tails) are particularly highly modified. These modifications include acetylation, methylation, ubiquitylation, phosphorylation and sumoylation. Acetylation is the most highly studied of these modifications. For example, acetylation of the K14 and K9 lysines of the tail of histone H3 by histone acetyltransferase enzymes (HATs) is generally correlated with transcriptional competence.

One mode of thinking is that this tendency of acetylation to be associated with "active" transcription is biophysical in nature. Because it normally has a positively charged nitrogen at its end, lysine can bind the negatively charged phosphates of the DNA backbone. The acetylation event converts the positively charged amine group on the side chain into a neutral amide linkage. This removes the positive charge, thus loosening the DNA from the histone. When this occurs, complexes like SWI/SNF and other transcriptional factors can bind to the DNA and allow transcription to occur. This is the "cis" model of epigenetic function. In other words, changes to the histone tails have a direct affect on the DNA itself.

Another model of epigenetic function is the "trans" model. In this model changes to the histone tails act indirectly on the DNA. For example, lysine acetylation may create a binding site for chromatin modifying enzymes (and basal transcription machinery as well). This Chromatin Remodeler can then cause changes to the state of the chromatin. Indeed, the bromodomain — a protein segment (domain) that specifically binds acetyl-lysine — is found in many enzymes that help activate transcription, including the SWI/SNF complex (on the protein polybromo). It may be that acetylation acts in this and the previous way to aid in transcriptional activation.

The idea that modifications act as docking modules for related factors is borne out by histone methylation as well. Methylation of lysine 9 of histone H3 has long been associated with constitutively transcriptionally silent chromatin (constitutive heterochromatin). It has been determined that a chromodomain (a domain that specifically binds methyl-lysine) in the transcriptionally repressive protein HP1 recruits HP1 to K9 methylated regions. One example that seems to refute this biophysical model for acetylation is that tri-methylation of histone H3 at lysine 4 is strongly associated with (and required for full) transcriptional activation. Tri-methylation in this case would introduce a fixed positive charge on the tail.

It has been shown that the histone lysine methyltransferase (KMT) is responsible for this methylation activity in the pattern of histones H3 & H4. This enzyme utilizes a catalytically active site called the SET domain (Supressor of variegation, Enhancer of zeste, Trithorax). The SET domain is a 130-amino acid sequence involved in modulating gene activities. This domain has been demonstrated to bind to the histone tail and causes the methylation of the histone.

Differing histone modifications are likely to function in differing ways; acetylation at one position is likely to function differently than acetylation at another position. Also, multiple modifications may occur at the same time, and these modifications may work together to change the behaviour of the nucleosome. The idea that multiple dynamic modifications regulate gene transcription in a systematic and reproducible way is called the histone code.

DNA methylation frequently occurs in repeated sequences, and helps to suppress the expression and mobility of 'transposable elements': Because 5-methylcytosine is chemically very similar to thymidine, CpG sites are frequently mutated and become rare in the genome, except at CpG islands where they remain unmethylated. Epigenetic changes of this type thus have the potential to direct increased frequencies of permanent genetic mutation. DNA methylation patterns are known to be established and modified in response to environmental factors by a complex interplay of at least three independent DNA methyltransferases, DNMT1, DNMT3A and DNMT3B, the loss of any of which is lethal in mice. DNMT1 is the most abundant methyltransferase in somatic cells, localizes to replication foci, has a 10–40-fold preference for hemimethylated DNA and interacts with the proliferating cell nuclear antigen (PCNA). By preferentially modifying hemimethylated DNA, DNMT1 transfers patterns of methylation to a newly synthesized strand after DNA replication, and therefore is often referred to as the 'maintenance' methyltransferase. DNMT1 is essential for proper embryonic development, imprinting and X-inactivation.

Histones H3 and H4 can also be manipulated through demethylation using histone lysine demethylase (KDM). This recently identified enzyme has a catalytically active site called the Jumonji domain (JmjC). The demethylation occurs when JmjC utilizes multiple cofactors to hydroxylate the methyl group, thereby removing it. JmjC is capable of demethylating mono-, di-, and tri-methylated substrates..

Chromosomal regions can adopt stable and heritable alternative states resulting in bistable gene expression without changes to the DNA sequence. Epigenetic control is often associated with alternative covalent modifications of histones. The stability and heritability of states of larger chromosomal regions are often thought to involve positive feedback where modified nucleosomes recruit enzymes that similarly modify nearby nucleosomes. A simplified stochastic model for this type of epigenetics is found here.

Because DNA methylation and chromatin remodeling play such a central role in many types of epigenic inheritance, the word "epigenetics" is sometimes used as a synonym for these processes. However, this can be misleading. Chromatin remodeling is not always inherited, and not all epigenetic inheritance involves chromatin remodeling.

It has been suggested that the histone code could be mediated by the effect of small RNAs. The recent discovery and characterization of a vast array of small (21- to 26-nt), non-coding RNAs suggests that there is an RNA component, possibly involved in epigenetic gene regulation. Small interfering RNAs can modulate transcriptional gene expression via epigenetic modulation of targeted promoters.

RNA Transcripts and their Encoded Proteins

Sometimes a gene, after being turned on, transcribes a product that (either directly or indirectly) maintains the activity of that gene. For example, Hnf4 and MyoD enhance the transcription of many liver- and muscle-specific genes, respectively, including their own, through the transcription factor activity of the proteins they encode. RNA signalling includes differential recruitment of a hierarchy of generic chromatin modifying complexes and DNA methyltransferases to specific loci by RNAs during differentiation and development.

Other epigenetic changes are mediated by the production of different splice forms of RNA, or by formation of double-stranded RNA (RNAi). Descendants of the cell in which the gene was turned on will inherit this activity, even if the original stimulus for gene-activation is no longer present. These genes are most often turned on or off by signal transduction, although in some systems where syncytia or gap junctions are important, RNA may spread directly to other cells or nuclei by diffusion. A large amount of RNA and protein is contributed to the zygote by the mother during oogenesis or via nurse cells, resulting in maternal effect phenotypes. A smaller quantity of sperm RNA is transmitted from the father, but there is recent evidence that this epigenetic information can lead to visible changes in several generations of offspring.

Prions

Prions are infectious forms of proteins. Proteins generally fold into discrete units which perform distinct cellular functions, but some proteins are also capable of forming an infectious conformational state known as a prion. Although often viewed in the context of infectious disease, prions are more loosely defined by their ability to catalytically convert other native state versions of the same protein to an infectious conformational state. It is in this latter sense that they can be viewed as epigenetic agents capable of inducing a phenotypic change without a modification of the genome.

Fungal prions are considered epigenetic because the infectious phenotype caused by the prion can be inherited without modification of the genome. PSI+ and URE3, discovered in yeast in 1965 and 1971, are the two best studied of this type of prion. Prions can have a phenotypic effect through the sequestration of protein in aggregates, thereby reducing that protein's activity. In PSI+ cells, the loss of the Sup35 protein (which is involved in termination of translation) causes ribosomes to have a higher rate of read-through of stop codons, an effect which results in suppression of nonsense mutations in other genes. The ability of Sup35 to form prions may be a conserved trait. It could confer an adaptive advantage by giving cells the ability to switch into a PSI+ state and express dormant genetic features normally terminated by premature stop codon mutations.

Structural Inheritance Systems

In ciliates such as *Tetrahymena* and *Paramecium*, genetically identical cells show heritable differences in the patterns of ciliary rows on their cell surface. Experimentally altered patterns can be transmitted to daughter cells. It seems existing structures act as templates for new structures. The mechanisms of such inheritance are unclear, but reasons exist to assume that multicellular organisms also use existing cell structures to assemble new ones.

Functions and Consequences

Development: Somatic epigenetic inheritance, particularly through DNA methylation and chromatin remodeling, is very important in the development of multicellular eukaryotic organisms. The genome sequence is static (with some notable exceptions), but cells differentiate into many different types, which perform different functions, and respond differently to the environment and intercellular signalling. Thus, as individuals develop, morphogens activate or silence genes in an epigenetically heritable fashion, giving cells a "memory".

In mammals, most cells terminally differentiate, with only stem cells retaining the ability to differentiate into several cell types ("totipotency" and "multipotency"). In mammals, some stem cells continue producing new differentiated cells throughout life, but mammals are not able to respond to loss of some tissues, for example, the inability to regenerate limbs, which some other animals are capable of. Unlike animals, plant cells do not terminally differentiate, remaining totipotent with the ability to give rise to a new individual plant. While plants do utilise many of the same epigenetic mechanisms as animals, such as chromatin remodeling, it has been hypothesised that plant cells do not have "memories", resetting their gene expression patterns at each cell division using positional information from the environment and surrounding cells to determine their fate.

Medicine

Epigenetics has many and varied potential medical applications. Congenital genetic disease is well understood, and it is also clear that epigenetics can play a role, for example, in the case of Angelman syndrome and Prader-Willi syndrome. These are normal genetic diseases caused by gene deletions or inactivation of the genes, but are unusually common because individuals are essentially hemizygous because of genomic imprinting, and therefore a single gene knock out is sufficient to cause the disease, where most cases would require both copies to be knocked out.

Evolution

Although epigenetics in multicellular organisms is generally thought to be a mechanism involved in differentiation, with epigenetic patterns "reset" when organisms reproduce, there have been some observations of transgenerational epigenetic inheritance (e.g., the phenomenon of paramutation observed in maize). Although most of these multigenerational epigenetic traits are gradually lost over several generations, the possibility remains that multigenerational epigenetics could be another aspect to evolution and adaptation. A sequestered germ line or Weismann barrier is specific to animals, and epigenetic inheritance is expected to be far more common in plants and microbes. These effects may require enhancements to the standard conceptual framework of the modern evolutionary synthesis.

Epigenetic features may play a role in short-term adaptation of species by allowing for reversible phenotype variability. The modification of epigenetic features associated with a region of DNA allows organisms, on a multigenerational time scale, to switch between phenotypes that express and repress that particular gene. When the DNA sequence of the region is not mutated, this change is reversible. It has also been speculated that organisms may take advantage of differential mutation rates associated with epigenetic features to control the mutation rates of particular genes. Interestingly, recent analysis have suggested that members of the APOBEC family of cytosine deaminases are capable of simultaneously mediating genetic and epigenetic inheritance using similar molecular mechanisms.

Epigenetic changes have also been observed to occur in response to environmental exposure—for example, mice given some dietary supplements have epigenetic changes affecting expression of the agouti gene, which affects their fur colour, weight, and propensity to develop cancer.

More than 100 cases of transgenerational epigenetic inheritance phenomena have been reported in a wide range of organisms, including prokaryotes, plants, and animals.

Epigenetic Effects in Humans

Genomic Imprinting and Related Disorders: Some human disorders are associated with genomic imprinting, a phenomenon in mammals where the father and mother contribute different epigenetic patterns for specific genomic loci in their germ cells. The best-known case of imprinting in human disorders is that of Angelman syndrome and Prader-Willi syndrome—both can be produced by the same genetic mutation, chromosome 15q partial deletion, and the particular syndrome that will develop depends on whether the mutation is inherited from the child's mother or from their father. This is due to the presence of genomic imprinting in the region. Beckwith-Wiedemann syndrome is also associated with genomic imprinting, often caused by abnormalities in maternal genomic imprinting of a region on chromosome 11.

Transgenerational Epigenetic Observations

Marcus Pembrey and colleagues also observed in the Överkalix study that the paternal (but not maternal) grandsons of Swedish boys who were exposed during preadolescence to famine in the 19th century were less likely to die of cardiovascular disease; if food was plentiful then diabetes mortality in the grandchildren increased, suggesting that this was a transgenerational epigenetic inheritance. The opposite effect was observed for females—the paternal (but not maternal) granddaughters of women who experienced famine while in the womb (and therefore while their eggs were being formed) lived shorter lives on average.

Cancer and Developmental Abnormalities

A variety of compounds are considered as epigenetic carcinogens—they result in an increased incidence of tumors, but they do not show mutagen activity (toxic compounds or pathogens that cause tumors incident to increased regeneration should also be excluded). Examples include diethylstilbestrol, arsenite, hexachlorobenzene, and nickel compounds.

Many teratogens exert specific effects on the fetus by epigenetic mechanisms. While epigenetic effects may preserve the effect of a teratogen such as diethylstilbestrol throughout the life of an affected child, the possibility of birth defects resulting from exposure of fathers or in second and succeeding generations of offspring has generally been rejected on theoretical grounds and for lack of evidence. However, a range of male-mediated abnormalities have been demonstrated, and more are likely to exist. FDA label information for Vidaza(tm), a formulation of 5-azacitidine (an unmethylatable analog of cytidine that causes hypomethylation when incorporated into DNA) states that "men should be advised not to father a child" while using the drug, citing evidence in treated male mice of reduced fertility, increased embryo loss, and abnormal embryo development. In rats, endocrine differences were observed in offspring of males exposed to morphine. In mice, second generation effects of diethylstilbesterol have been described occurring by epigenetic mechanisms.

Recent studies have shown that the Mixed Lineage Leukemia (MLL) gene causes leukemia by rearranging and fusing with other genes in different chromosomes, which is a process under epigenetic control.

Other investigations have concluded that alterations in histone acetylation and DNA methylation occur in various genes influencing prostate cancer.

In 2008, the National Institutes of Health announced that $190 million had been earmarked for epigenetics research over the next five years. In announcing the funding, government officials noted that epigenetics has the potential to explain mechanisms of aging, human development, and the origins of cancer, heart disease, mental illness, as well as several other conditions. Some investigators, like Randy Jirtle, PhD, of Duke University Medical Centre, think epigenetics may ultimately turn out to have a greater role in disease than genetics.

DNA Methylation in Cancer

DNA methylation is an important regulator of gene transcription and a large body of evidence has demonstrated that aberrant DNA methylation is associated with unscheduled gene silencing, and the genes with high levels of 5-methylcytosine in their promoter region are transcriptionally silent. DNA methylation is essential during embryonic development, and in somatic cells, patterns of DNA methylation are generally transmitted to daughter cells with a high fidelity.

Aberrant DNA methylation patterns have been associated with a large number of human malignancies and found in two distinct forms: hypermethylation and hypomethylation compared to normal tissue. Hypermethylation is one of the major epigenetic modifications that repress transcription via promoter region of tumour suppressor genes. Hypermethylation typically occurs at CpG islands in the promoter region and is associated with gene inactivation. Global hypomethylation has also been implicated in the development and progression of cancer through different mechanisms.

Variant Histones H2A in Cancer

The histone variants of the H2A family are highly conserved in mammals, playing critical roles in regulating many nuclear processes by altering chromatin structure. One of the key H2A variants, H2A.X, marks DNA damage, facilitating the recruitment of DNA repair proteins to restore genomic integrity. Another variant, H2A.Z, plays an important role in both gene activation and repression. A high level of H2A.Z expression is ubiquitously detected in many cancers and is significantly associated with cellular proliferation and genomic instability.

Cancer Treatment

Current research has shown that epigenetic pharmaceuticals could be a putative replacement or adjuvant therapy for currently accepted treatment methods such as radiation and chemotherapy, or could enhance the effects of these current treatments. It has been shown that the epigenetic control of the proto-onco regions and the tumour suppressor sequences by conformational changes in histones directly affects the formation and progression of cancer Epigenetics also has the factor of reversibility, a characteristic that other cancer treatments do not offer.

Drug development has mainly focused on Histone Acetyltransferase (HAT) and Histone Deactylase (HDAC), including the introduction of the new pharmaceutical Vorinostat, a HDAC inhibitor, to the market. HDAC specifically has been shown to play an integral role in the progression of oral squamous cancer.

Current front-runner candidates for new drug targets are Histone Lysine Methyltransferases (KMT) and Protein Arginine Methyltransferases (PRMT).

Twin Studies

Recent studies involving both dizygotic and monozygotic twins have produced some evidence of epigenetic influence in humans.

Epigenetics in Microorganisms

Bacteria make widespread use of postreplicative DNA methylation for the epigenetic control of DNA-protein interactions. Bacteria make use of DNA adenine methylation (rather than DNA cytosine methylation) as an epigenetic signal. DNA adenine methylation is important in bacteria virulence in organisms such as *Escherichia coli, Salmonella, Vibrio, Yersinia, Haemophilus,* and *Brucella.* In *Alphaproteobacteria*, methylation of adenine regulates the cell cycle and couples gene transcription to DNA replication. In *Gammaproteobacteria*, adenine methylation provides signals for DNA replication, chromosome segregation, mismatch repair, packaging of bacteriophage, transposase activity and regulation of gene expression.

The filamentous fungus *Neurospora crassa* is a prominent model system for understanding the control and function of cytosine methylation. In this organisms, DNA methylation is associated with relics of a genome defense system called RIP (repeat-induced point mutation) and silences gene expression by inhibiting transcription elongation.

The yeast prion PSI is generated by a conformational change of a translation termination factor, which is then inherited by daughter cells. This can provide a survival advantage under adverse conditions. This is an example of epigenetic regulation enabling unicellular organisms to respond rapidly to environmental stress. Prions can be viewed as epigenetic agents capable of inducing a phenotypic change without modification of the genome.

Modification of DNA

In eukaryotes, the accessibility of large regions of DNA can depend on its chromatin structure, which can be altered as a result of histone modifications directed by DNA methylation, ncRNA, or DNA-binding protein.

Chemical

Methylation of DNA is a common method of gene silencing. DNA is typically methylated by methyltransferase enzymes on cytosine nucleotides in a CpG dinucleotide sequence (also called "CpG islands" when densely clustered). Analysis of the pattern of methylation in a given region of DNA (which can be a promoter) can be achieved through a method called bisulphite mapping. Methylated cytosine residues are unchanged by the treatment, whereas unmethylated ones are changed to uracil. The differences are analysed by DNA sequencing or by methods developed to quantify SNPs, such as Pyrosequencing (Biotage) or MassArray (Sequenom), measuring the relative amounts of C/T at the CG dinucleotide. Abnormal methylation patterns are thought to be involved in oncogenesis.

Structural

Transcription of DNA is dictated by its structure. In general, the density of its packing is indicative of the frequency of transcription. Octameric protein complexes called nucleosomes are responsible for the amount of supercoiling of DNA, and these complexes can be temporarily modified by processes such as phosphorylation or more permanently modified by processes such as methylation. Such modifications are considered to be responsible for more or less permanent changes in gene expression levels.

Histone acetylation is also an important process in transcription. Histone acetyltransferase enzymes (HATs) such as CREB-binding protein also dissociate the DNA from the histone complex, allowing transcription to proceed. Often, DNA methylation and histone deacetylation work together in gene silencing. The combination of the two seems to be a signal for DNA to be packed more densely, lowering gene expression.

Regulation of Transcription

Regulation of transcription controls when transcription occurs and how much RNA is created. Transcription of a gene by RNA polymerase can be regulated by at least five mechanisms:

- Specificity factors alter the specificity of RNA polymerase for a given promoter or set of promoters, making it more or less likely to bind to them (i.e., sigma factors used in prokaryotic transcription).

- Repressors bind to non-coding sequences on the DNA strand that are close to or overlapping the promoter region, impeding RNA polymerase's progress along the strand, thus impeding the expression of the gene.

- General transcription factors position RNA polymerase at the start of a protein-coding sequence and then release the polymerase to transcribe the mRNA.

- Activators enhance the interaction between RNA polymerase and a particular promoter, encouraging the expression of the gene. Activators do this by increasing the attraction of RNA polymerase for the promoter, through interactions with subunits of the RNA polymerase or indirectly by changing the structure of the DNA.

- Enhancers are sites on the DNA helix that are bound to by activators in order to loop the DNA bringing a specific promoter to the initiation complex. Enhancers are much more common in eukaryote than prokaryotes, where only a few examples exist (to date).

Post-transcriptional Regulation

After the DNA is transcribed and mRNA is formed, there must be some sort of regulation on how much the mRNA is translated into proteins. Cells do this by modulating the capping, splicing, addition of a Poly(A) Tail, the sequence-specific nuclear export rates, and, in several contexts, sequestration of the RNA transcript. These processes occur in eukaryotes but not in prokaryotes. This modulation is a result of a protein or transcript that, in turn, is regulated and may have an affinity for certain sequences.

Regulation of Translation

The translation of mRNA can also be controlled by a number of mechanisms, mostly at the level of initiation. Recruitment of the small ribosomal subunit can indeed be modulated by mRNA secondary structure, antisense RNA binding, or protein binding.

In both prokaryotes and eukaryotes, a large number of RNA binding proteins exist, which often are directed to their target sequence by the secondary structure of the transcript, which may change depending on certain conditions, such as temperature or presence of a ligand (aptamer). Some transcripts act as ribozymes and self-regulate their expression.

Examples of Gene Regulation

- Enzyme induction is a process in which a molecule (e.g., a drug) induces (i.e., initiates or enhances) the expression of an enzyme.
- The induction of heat shock proteins in the fruit fly *Drosophila melanogaster*.
- The Lac operon is an interesting example of how gene expression can be regulated.
- Viruses, despite having only a few genes, possess mechanisms to regulate their gene expression, typically into an early and late phase, using collinear systems regulated by anti-terminators (lambda phage) or splicing modulators (HIV).

Developmental Biology

A large number of studied regulatory systems come from developmental biology. Examples include:

- The colinearity of the Hox gene cluster with their nested antero-posterior patterning
- It has been speculated that pattern generation of the hand (digits - interdigits) The gradient of Sonic hedgehog (secreted inducing factor) from the zone of polarizing activity in the limb, which creates a gradient of active Gli3, which activates Gremlin, which inhibits BMPs also secreted in the limb,

resulting in the formation of an alternating pattern of activity as a result of this reaction-diffusion system.

- Somitogenesis is the creation of segments (somites) from a uniform tissue (Pre-somitic Mesoderm, PSM). They are formed sequentially from anterior to posterior. This is achieved in amniotes possibly by means of two opposing gradients, Retinoic acid in the anterior (wavefront) and Wnt and Fgf in the posterior, coupled to an oscillating pattern (segmentation clock) composed of FGF + Notch and Wnt in antiphase.
- Sex determination in the soma of a Drosophila requires the sensing of the ratio of autosomal genes to sex chromosome-encoded genes, which results in the production of sexless splicing factor in females, resulting in the female isoform of doublesex.

Circuitry

Up-regulation and Down-regulation

Up-regulation is a process that occurs within a cell triggered by a signal (originating internal or external to the cell), which results in increased expression of one or more genes and as a result the protein(s) encoded by those genes. On the converse, down-regulation is a process resulting in decreased gene and corresponding protein expression.

- Up-regulation occurs, for example, when a cell is deficient in some kind of receptor. In this case, more receptor protein is synthesized and transported to the membrane of the cell and, thus, the sensitivity of the cell is brought back to normal, reestablishing homeostasis.
- Down-regulation occurs, for example, when a cell is overstimulated by a neurotransmitter, hormone, or drug for a prolonged period of time, and the expression of the receptor protein is decreased in order to protect the cell.

Inducible vs. Repressible Systems

Gene Regulation can be summarized as how they respond:

- Inducible systems - An inducible system is off unless there is the presence of some molecule (called an inducer) that allows for gene expression. The molecule is said to "induce expression". The manner by which this happens is dependent on the control mechanisms as well as differences between prokaryotic and eukaryotic cells.

- Repressible systems - A repressible system is on except in the presence of some molecule (called a corepressor) that suppresses gene expression. The molecule is said to "repress expression". The manner by which this happens is dependent on the control mechanisms as well as differences between prokaryotic and eukaryotic cells.

Theoretical Circuits

- Repressor/Inducer: an activation of a sensor results in the change of expression of a gene
- negative feedback: the gene product downregulates its own production directly or indirectly, which can result in
  - keeping transcript levels constant/proportional to a factor
  - inhibition of run-away reactions when coupled with a positive feedback loop
  - creating an oscillator by taking advantage in the time delay of transcription and translation, given that the mRNA and protein half-life is shorter
- positive feedback: the gene product upregulates its own production directly or indirectly, which can result in
  - signal amplification
  - bistable switches when two genes inhibit each other and have both positive feedback
  - pattern generation.

Methods

In general, most experiments investigating differential expression used whole cell extracts of RNA, called steady-state levels, to determine which genes changed and by how much they did.

These are, however, not informative of where the regulation has occurred and may actually mask conflicting regulatory processess, but it is still the most commonly analysed (QPCR and DNA microarray).

When studying gene expression, there are several methods to look at the various stages. In eukaryotes these include:

- The chromatin conformation of the region can be determined by ChIP-chip analysis by pulling down RNA Polymerase II, Histone 3 modifications, Trithorax-group protein, Polycomb-group protein, or any other DNA-binding element to which a good antibody is available.

- Epistatic interactions can be investigated by synthetic genetic array analysis

- Due to post-transcriptional regulation, transcription rates and total RNA levels differ significantly. To measure the transcription rates nuclear run-on assays can be done and newer high-throughput methods are being developed, using thiol labelling instead of radioactivity.

- Only 5% of the RNA polymerised in the nucleus actually exists, and not only introns, abortive products, and non-sense transcripts are degradated. Therefore, the differences in nuclear and cytoplasmic levels can be see by separating the two fractions by gentle lysis.

- Alternative splicing can be analysed with a splicing array or with a tiling array.

- All in vivo RNA is complexed as RNPs. The quantity of transcripts bound to specific protein can be also analysed by RIP-Chip. For example, DCP2 will give an indication of sequestered protein; ribosome-bound gives and indication of transcripts active in transcription (although it should be noted that a more dated method, called polysome fractionation, is still popular in some labs)

- Protein levels can be analysed by Mass spectrometry, which can be compared only to QPCR data, as microarray data is relative and not absolute.

- RNA and protein degradation rates are measured by means of transcription inhibitors (actinomycin D or α-amanitin) or translation inhibitors (Cycloheximide), respectively.

- Enhancer (genetics).

In genetics, an enhancer is a short region of DNA that can be bound with proteins (namely, the trans-acting factors, much like a set of transcription factors) to enhance transcription levels of genes (hence the name) in a gene cluster. While enhancers are usually cis-acting, an enhancer does not need to be particularly close to the genes it acts on, and need not be located on the same chromosome.

In eukaryotic cells the structure of the chromatin complex of DNA is folded in a way that functionally mimics the supercoiled state characteristic of prokaryotic DNA, so that although the enhancer DNA is far from the gene in regard to the number of nucleotides, it is geometrically close to the promoter and gene. This allows it to interact with the general transcription factors and RNA polymerase II. An enhancer may be located upstream or downstream of the gene that it regulates.

Furthermore, an enhancer does not need to be located near to the transcription initiation site to affect the transcription of a gene, as some have been found to bind several hundred thousand base pairs upstream or downstream of the start site. Enhancers do not act on the promoter region itself, but are bound by activator proteins. These activator proteins interact with the mediator complex, which recruits polymerase II and the general transcription factors which then begin transcribing the genes. Enhancers can also be found within introns. An enhancer's orientation may even be reversed without affecting its function. Additionally, an enhancer may be excised and inserted elsewhere in the chromosome, and still affect gene transcription. That is the reason that intron polymorphisms are checked though they are not translated.

Theories

Currently, there are two different theories on the information processing that occurs on enhancers:

- Enhanceosomes - rely on highly cooperative, coordinated action and can be disabled by single point mutations that move or remove the binding sites of individual proteins.
- Flexible billboards - less integrative, multiple proteins independently regulate gene expression and their sum is read in by the basal transcriptional machinery.

HACNS1

HACNS1 (also known as CENTG2 and located in the Human Accelerated Region 2) is a gene enhancer "that may have contributed to the evolution of the uniquely opposable human thumb, and possibly also modifications in the ankle or foot that allow humans to walk on two legs". Evidence to date shows that of the 110,000 gene enhancer sequences identified in the human genome, HACNS1 has undergone the most change during the evolution of humans following the split with the ancestors of chimpanzees.

Eubacteria - Gene Regulation and Protein Synthesis

Gene expression in many bacteria is regulated through the existence of operons. An operon is a cluster of genes whose protein products have related functions. For instance, the *lac* operon includes one gene that transports lactose sugar into the cell and another that breaks it into two parts. These genes are under the control of the same promoter, and so are transcribed and translated into protein at the same time. RNA polymerase can only reach the promoter if a repressor is not blocking it; the *lac* repressor is dislodged by lactose. In this way, the bacterium uses its resources to make lactose-digesting enzymes only when lactose is available.

Other genes are expressed constantly at low levels; their protein products are required for "housekeeping" functions such as membrane synthesis and DNA repair. One such enzyme is DNA gyrase, which relieves strain in the double helix during replication and repair. DNA gyrase is the target for the antibiotic ciproflaxin (sold under the name Cipro), effective against *Bacillus anthracis*, the cause of anthrax. Since eukaryotes do not have this type of DNA gyrase, they are not harmed by the action of this antibiotic.

As in eukaryotes, translation (protein synthesis) occurs on the ribosome. Without a nucleus to exclude it, the ribosome can attach to the messenger RNA even while the RNA is still attached to the DNA. Multiple ribosomes can attach to the same mRNA, making multiple copies of the same protein.

The ribosomes of eubacteria are similar in structure to those in eukaryotes and archaea, but differ in molecular detail. This has two important consequences. First, sequencing ribosomal RNA molecules is a useful tool for understanding the evolutionary diversification of the Eubacteria. Organisms with more similar sequences are presumed to be more closely related. The same tool has been used to show that Archaea and Eubacteria are not closely related, despite their outward similarities. Indeed, Archaea are more closely related to eukaryotes (including humans) than they are to Eubacteria.

Second, the differences between bacterial and eukaryotic ribosomes can be exploited in designing antibacterial therapies. Various unique parts of the bacterial ribosome are the targets for numerous antibiotics, including streptomycin, tetracycline, and erythromycin.

Gene Expression Regulates Cell Differentiation

All of the cells within a complex multicellular organism such as a human being contain the same DNA; however, the body of such an organism is clearly composed of many different types of cells. What, then, makes a liver cell different from a skin or muscle cell? The answer lies in the way each cell deploys its genome. In other words, the particular combination of genes that are turned on (expressed) or turned off (repressed) dictates cellular morphology (shape) and function. This process of gene expression is regulated by cues from both within and outside cells, and the interplay between these cues and the genome affects essentially all processes that occur during embryonic development and adult life.

Do all Cells Really Contain the Same DNA?

Several lines of evidence support the proposal that all of the cells within a multicellular organism contain the same genome. For instance, although you started as a single cell with a half-genome from each parent, that single cell quickly divided and new cells began to differentiate, or become different from each other. While this process of differentiation established a wide variety of cell types (e.g., skin, liver, muscle, etc.), it was not accompanied by any permanent loss of genetic material. This is demonstrated by the fact that fully differentiated cell types are still capable, within the right environment, of giving rise to an entire new animal. This capability was first shown by way of an experiment in which the nucleus of an adult frog skin cell was transplanted into an enucleated donor embryo, eventually leading to the development of a cloned adult frog (Gurdon *et al.*, 1975). Later, the intact complete genome of a differentiated cell was used in the cloning of the famous sheep Dolly.

Today, researchers understand that the specialized, differentiated cell types of the adult body contain a genome as complete as any embryo's. This fascinating demonstration has led to the proposal that changes in gene expression, rather than losses of genetic material, play a key role in guiding and maintaining cell differentiation.

Cell-Extrinsic Regulation of Gene Expression

Gene expression is regulated by factors both extrinsic and intrinsic to the cell. Cell-extrinsic factors that regulate expression include environmental cues, such as small molecules, secreted proteins, temperature, and oxygen. These cues can originate from other cells within the organism, or they can come from the organism's environment. Within the organism, cells communicate with each other by sending and receiving secreted proteins, also known as growth factors, morphogens, cytokines, or signaling molecules. Receipt of these signaling molecules triggers intercellular signaling cascades that ultimately cause semipermanent changes in transcription or expression of genes. Such changes in gene expression can include turning genes completely on or off, or just slightly tweaking the level of transcript produced. This process is thought to regulate a vast number of cell behaviours, including cell fate decisions during embryogenesis, cell function, and chemotaxis.

In addition, gene expression changes can lead to changes in an entire organism, such as molting in insects. In *Drosophila*, for example, the molting process is regulated by levels of a hormone called ecdysone. This hormone acts as a signal, triggering a cascade of events and leading to changes in gene expression. Not surprisingly, the genes that are expressed in response to ecdysone are also the genes that are involved in the molting process (White *et al.*, 1997). Thus, ecdysone acts on the organism level as a cell-extrinsic factor to bring about physiologically meaningful changes in gene expression. What is also interesting is that scientists can learn more about a physiological process like metamorphosis by studying how gene expression patterns change over time. For example, although researchers were aware that ecdysone results in a decrease of transcription from some loci, such as those involved in the glycolytic pathway, microarray data suggest that ecdysone-induced metamorphosis also downregulates genes involved with fatty acid oxidation, amino acid metabolism, oxidative phosphorylation, and other pathways. This suggests that there is a more global repression of metabolic activity during molting. Specifically, during metamorphosis, the larval muscle cells are degraded, and muscle-specific genes are downregulated. Simultaneously, the development of the nervous system begins, and the genes involved in neuronal differentiation are induced.

Cell-Intrinsic Regulation of Gene Expression

Although differentiation is not thought to occur by permanent loss of genetic material, DNA can be modified in a way that affects gene expression. For instance, DNA and its associated histone proteins (together known as chromatin) can be chemically modified by a cell's own machinery. Chromatin modification can affect gene expression by changing the accessibility of genes to transcription factors, in either a positive or a negative manner. Two major classes of such chemical modifications include DNA methylation and histone modification (methylation and/or acetylation). These changes are often described as epigenetic because they do not act to alter the primary DNA sequence but instead act at a level just above the DNA sequence. Although DNA methylation and histone modification are not genetic, cells have mechanisms to copy this epigenetic information during their division so that their daughter cells contain the same regulatory data.

Changes in chromatin modification play an important role in regulating gene expression during developmental cell-type specification as well. For example, chromatin-modifying proteins play an essential role in muscle cell differentiation via interactions with key muscle-promoting transcription factors MyoD and MEF. That is, these factors are thought to help recruit chromatin modifying factors, such as histone acetyltransferases and deacetylases. In so doing, MyoD and MEF alter access to their target sites upstream of muscle differentiation genes. For instance, MyoD binds histone acetyltransferases p300 and PCAF, and this activity is essential for muscle cell differentiation (Puri *et al.*, 1997). This example provides evidence for a link among chromatin modifications, transcription factors, and, ultimately, cell-fate-specific changes in gene expression.

Chromatin modification can be stable over the life of an organism, thereby effectively permanently influencing gene expression. However, that is not to say that chromatin modification is irreversible. For instance, chromatin can become mismodified in certain cancers (Vucic *et al.*, 2008), suggesting that, although important, the change is not permanent. Moreover, chromatin modifications are usually erased and reset during the production of gametes, such that the adult program of intrinsic cues is replaced with a program more suited to embryonic development.

In fact, embryonic cell types are known to contain a unique set of chromatin modifications that are different from those found in adult cell types. This has led to the tantalizing proposal that chromatin modification helps lock in changes in gene expression that are required during development.

The permanent silencing of the genes involved only in embryogenesis could then drive the development of cells towards more mature cell types. By blocking accessibility of transcription machinery, for example, chromatin modification could prevent the need for continued repression through active binding of a repressive transcription factor. Alternatively, the genes required for an adult cell type might contain chromatin modifications (especially histone acetylation) that cause the DNA to become open and, therefore, more accessible to the transcription machinery.

Interestingly, embryonic cell types have been found to contain a signature chromatin modification in the regions that regulate the expression of genes involved in early embryonic development. Such regions were found to contain chromatin modifications with both silencing and promoting characteristics. The finding of these bivalent (two-directional) markers in association with genes important for embryonic development has led to the belief that embryonic cells exist in a special epigenetic state, wherein they can choose to remain embryonic (as in an embryonic stem cell) or to differentiate (as in normal development), and bivalent domains provide a means by which to quickly choose between the two options.

Together, these lines of evidence have led to an emerging hypothesis that cell-cell signaling and epigenetic changes converge to guide cell differentiation decisions both during development and beyond.

Transcription Factors and Transcriptional Control in Eukaryotic Cells

Do complex organisms have more genes than simpler organisms? Now that researchers can sequence whole genomes and have done so for a number of organisms, they know that many vertebrates have only about twice as many genes as invertebrates, and many of these are the result of duplication of existing genes rather than development of new ones. But if there are not that many new genes, what is responsible for the incredible diversity in plant and animal species?

The simple answer to this question is that eukaryotes have developed a more complex way of controlling expression of their existing genes than prokaryotes. This system of expression control relies on a group of proteins known as transcription factors (TFs), and it allows eukaryotes to alter their cell types and growth patterns in a variety of ways. TFs are not solely responsible for gene regulation; eukaryotes also rely on cell signaling, RNA splicing, siRNA control mechanisms, and chromatin modifications. However, TFs that bind to cis-regulator DNA sequences are responsible for either positively or negatively influencing the transcription of specific genes, essentially determining whether a particular gene will be turned "on" or "off" in an organism.

Transcription Factors Recognize Specific DNA Sequences

Much of the complexity in differentiation in animal and plant cells can be attributed to the evolution of elaborate systems made up of short (6 to 8 base pair) *cis*-regulatory DNA sequences or motifs, as well as the TFs that bind to the motifs, interact with each other to form complexes, and recruit RNA polymerase II (Levine & Tjian, 2003). Most eukaryotic genes have promoters that consist of the TATA box close to the 5' end of the gene and, farther upstream, several motifs recognized by specific transcription factors.

In addition, many genes have one or more other nearby sequences called enhancers. Enhancers affect transcription; these sequences occur upstream, downstream, or within introns, and they continue to work whether in the normal orientation or turned backward in the genome. In yeast, no enhancers are known; instead, there are only upstream activator sequences (UASs). Enhancers can be found thousands of base pairs from a promoter, whereas UASs are generally within a few hundred base pairs upstream. Typical RNA polymerase II promoters can be influenced by many enhancers and by multiple factors bound to the promoter and enhancer sequences.

The mode of action of TFs is to recognize and bind to a segment of DNA in the promoter and/or enhancer region. Often, a change in the conformation, or three-dimensional structure of a TF, will accompany DNA binding. For example, the two loops in NFATC1 that interact with DNA are found in different conformations, depending on whether NFATC1 is complexed with DNA or not. Moreover, the structure of different TF families, described later in this article, results in specific areas in these protein complexes that interact with the DNA recognition motif. The recognition motif is usually only about 6 to 10 base pairs long.

Experiments have shown that TFs can bind tightly, both within cells and *in vitro*. After TFs bind to promoter or enhancer regions of the DNA, they interact with other bound TFs and recruit RNA polymerase II. Their influence, however, can be either positive or negative, depending on the presence of other functional domains on the protein and the overall impact of the entire TF complex. A typical TF has multiple functional domains, not only for recognizing and binding to the appropriate DNA strand, but also for interactions with other TFs, with proteins called coactivators, with RNA polymerase II, with chromatin remodeling complexes, and with small noncoding RNAs.

TFs control many important parts of development; therefore, organisms with a deletion of a TF gene exhibit profound irregularities in organization and development. For example, in *Drosophila*, deletion of the TF antennapedia gene results in the development of the antennal imaginal disc into legs rather than antennae.

Transcription Factors Exert Combinatorial Control

Many TFs are known to facilitate transcription at hundreds of different promoters, while some are only active at a select few. Laboratory techniques such as chromatin immunoprecipitation (ChIP) and DNA microarrays are commonly used to study the target DNA motifs recognized by individual TFs (Iyer *et al.*, 2001). Signal molecules can influence activation by TFs by covalently binding or modifying their functional domains. It is even possible for a TF to respond to a physical signal, such as red or far-red light, but the signal must be transduced to a chemically modified activator that interacts with the TF.

The complexity and fine gradations of DNA expression in eukaryotes result from combinatorics, in that the combination of chromatin and TF signals, rather than the individual TF signal, is read out. Thus, transcriptional control is dependent on the interactions of all the TFs and whether they attract RNA polymerase or block it from initiating transcription. Multiple TFs can accumulate, creating a bulk the size of a ribosome. Once bound together, changes to the functional domains of a TF and/or covalent interactions with other factors can turn transcription on or off, depending on whether they allow or prohibit the recruitment of RNA polymerase.

A typical enhancer can be up to 500 base pairs in length and contain multiple binding sites for at least two or three different TFs (Levine & Tijan, 2003). Two TFs bound at sites near one another on the DNA strand can combine to form a dimer and bend the DNA in what is believed to be part of the activation process. Chromatin structure allows activators to associate with one another, even when they are bound to DNA sequences many hundreds of base pairs apart. Some TFs are believed to act as tethering elements between distant enhancers and promoters by forming connections with other proteins.

The Evolution of Transcription Factor Families

Higher organisms have a large number of diverse TF families defined by the sequence of their DNA-binding domains. Evolutionary studies have shown that although the DNA-binding motif is highly conserved among plants and animals, the remainder of these organisms' protein sequences is often very different. In addition, a particular TF family may have different roles in plants than in animals, and some new TFs have evolved in each kingdom since their divergence.

In many animals, including humans, a prominent group of genes involved in cell development, including many that encode TFs, contain a 180 base-pair sequence called the homeobox. The homeobox encodes a 60-amino acid protein segment called the homeodomain, which recognizes and binds to promoters in the DNA of its target genes. Complete control over transcription, and sometimes binding, is dependent on interactions between TFs, so activation often depends upon the presence of another TF. A similar system of gene recognition is found in plants, where the DNA-binding domain is called the MADS box.

TFs often have certain specific DNA-binding motifs, a common one being the basic helix-loop-helix (bHLH) structure that recognizes a specific sequence of DNA and sits on the DNA like a train car on a track. One such example is the TF MyoD (myoblast determination). Expression of the *MyoD* gene results in production of MyoD protein, which binds to the promoters of muscle-determining genes, causing the differentiation of muscle precursor cells (myoblasts) into muscle fibres. MyoD also binds to its own promoter, thus maintaining its own levels in differentiated muscle cells and their progeny.

In addition to bHLH, there are some other common structural motifs for recognition and binding of DNA, and these are found in most regulatory proteins. These are the helix-turn-helix, zinc finger, and leucine zipper. Proteins having each of these motifs are effective because they fit neatly into the major or minor grooves of the DNA strand, and also because they expose specific amino acids at the appropriate places to form hydrogen bonds with the nucleotide bases. Molecular genetic techniques can be used to change any amino acids to test whether this affects the binding affinity of the TF for the target.

Complexity and Transcription Factors

Complexity of transcriptional control can be illustrated by comparing the number and locations of *cis*-control elements in higher and lower eukaryotes. For instance, *Drosophila* typically has several enhancers for a single gene of 2 to 3 kilobases, scattered over a large (10 kilobase) region of DNA, while, as described earlier, yeast have no enhancers but instead use one UAS sequence per gene, located upstream. Long-range regulation is thought to be indicative of the need for a higher level of control over genes involved in cell development and differentiation.

The yeast genome encodes around 300 TFs, or one per every 20 genes, while humans express approximately 3,000 TFs, or one per every 10 genes. With combinatorial control, the twofold increase in TFs per gene actually translates into many more possible combinations of interactions, allowing for the dramatic increase in diversity among organisms. When we consider the additional complexities of chromatin remodeling, regulated mRNA stability, and translational control, it is easier to understand how the cells of higher organisms can produce such an enormous variety of genetic responses to environmental signals.

Nucleic Acids to Amino Acids: DNA Specifies Protein

Once it was determined that messenger RNA (mRNA) serves as a copy of chromosomal DNA and specifies the sequence of amino acids in proteins, the question of how this process is actually carried out naturally followed. It had long been known that only 20 amino acids occur in naturally derived proteins. It was also known that there are only four nucleotides in mRNA: adenine (A), uracil (U), guanine (G), and cytosine (C). Thus, 20 amino acids are coded by only four unique bases in mRNA, but just how is this coding achieved?

The Codon

The discordance between the number of nucleic acid bases and the number of amino acids immediately eliminates the possibility of a code of one base per amino acid. In fact, even two nucleotides per amino acid (a doublet code) could not account for 20 amino acids (with four bases and a doublet code, there would only be 16 possible combinations [$4^2 = 16$]). Thus, the smallest combination of four bases that could encode all 20 amino acids would be a triplet code. However, a triplet code produces 64 ($4^3 = 64$) possible combinations, or codons. Thus, a triplet code introduces the problem of there being more than three times the number of codons than amino acids. Either these "extra" codons produce redundancy, with multiple codons encoding the same amino acid, or there must instead be numerous dead-end codons that are not linked to any amino acid.

Preliminary evidence indicating that the genetic code was indeed a triplet code came from an experiment by Francis Crick and Sydney Brenner (1961). This experiment examined the effect of frameshift mutations on protein synthesis. Frameshift mutations are much more disruptive to the genetic code than simple base substitutions, because they involve a base insertion or deletion, thus changing the number of bases and their positions in a gene. For example, the mutagen proflavine causes frameshift mutations by inserting itself between DNA bases. The presence of proflavine in a DNA molecule thus interferes with the molecule's replication such that the resultant DNA copy has a base inserted or deleted.

Crick and Brenner showed that proflavine-mutated bacteriophages (viruses that infect bacteria) with single-base insertion or deletion mutations did not produce functional copies of the protein encoded by the mutated gene. The production of defective proteins under these circumstances can be attributed to misdirected translation. Mutant proteins with two- or four-nucleotide insertions or deletions were also nonfunctional. However, some mutant strains became functional again when they accumulated a total of three extra nucleotides or when they were missing three nucleotides. This rescue effect provided compelling evidence that the genetic code for one amino acid is indeed a three-base, or triplet, code.

Decoding the Genetic Code

Once the budding molecular biology community was convinced about the triplet code, the race to decode which triplets specified which amino acids began. The simplest way to decipher the code would be to start with an mRNA molecule of known sequence, use it to direct the synthesis of a protein, and then determine the amino acid sequence of the synthesized protein. Then, comparison of the original mRNA sequence with the amino acid sequence of the synthesized protein could provide a means for directly decoding the genetic code.

In further experiments to decode the other codons, Nirenberg and his colleagues made artificial RNAs containing defined proportions of two or three different bases. As previously mentioned, polynucleotide phosphorylase joins nucleotides randomly; as a result, these artificial RNAs contained random mixtures of the bases in proportion to the amounts of bases mixed. Hence, the resulting products provided clues that the researchers could use to deduce potential codon–amino acid relationships.

For example, when A and C were mixed with polynucleotide phosphorylase, the resulting RNA molecules contained eight different triplet codons: AAA, AAC, ACC, ACA, CAA, CCA, CAC, and CCC.

These eight random poly(AC) RNAs produced proteins containing only six amino acids: asparagine, glutamine, histidine, lysine, proline, and threonine. Remember that previous experiments had already revealed that CCC and AAA code for proline and lysine, respectively. Thus, the four newly incorporated amino acids could only be encoded by AAC, ACC, ACA, CAA, CCA, and/or CAC. With the random sequence approach, the decoding endeavor was almost completed, but some work remained to be done.

Thus, in 1965, H. Gobind Khorana and his colleagues used another method to further crack the genetic code. These researchers had the insight to employ chemically synthesized RNA molecules of known repeating sequences rather than random sequences. For example, an artificial mRNA of alternating guanine and uracil nucleotides (GUGUGUGUGUGU) should be read in translation as two alternating codons, GUG and UGU, thus encoding a protein of two alternating amino acids. Translation of the artificial GUGU mRNA yielded a protein of alternating cysteine and valine residues. However, this technique alone could not determine whether GUG or UGU encoded cysteine, for example.

Next, Nirenberg and Philip Leder developed a technique using ribosome-bound transfer RNAs (tRNAs). They showed that a short mRNA sequence—even a single codon (three bases)—could still bind to a ribosome, even if this short sequence was incapable of directing protein synthesis. The ribosome-bound codon could then base pair with a particular tRNA that carried the amino acid specified by the codon.

Nirenberg and Leder thus synthesized many short mRNAs with known codons. They then added the mRNAs one by one to a mix of ribosomes and aminoacyl-tRNAs with one amino acid radioactively labeled. For each, they determined whether the aminoacyl-tRNA was bound to the short mRNA-like sequence and ribosome (the rest passed through the filter), providing conclusive demonstrations of the particular aminoacyl-tRNA that bound to each mRNA codon.

Degeneracy of the Amino Acid Code

Examination of the full table of codons enables one to immediately determine whether the "extra" codons are associated with redundancy or dead-end codes. Note that both possibilities occur in the code. There are only a few instances in which one codon codes for one amino acid, such as the codon for tryptophan. Note also that the codon for the amino acid methionine (AUG) acts as the start signal for protein synthesis in an mRNA. Moreover, the genetic code also includes stop codons, which do not code for any amino acid. The stop codons serve as termination signals for translation. When a ribosome reaches a stop codon, translation stops, and the polypeptide is released.

Myofilaments

Morphologically, there are two kinds of myofilaments, *thick myofilaments* that are about 1.5 μm long and 10 nm wide and that are separated by a 40 nm space, and *thin myofilaments* that are about 1.0 μm long and 5 nm in diameter. As will be mentioned later, thick myofilaments are made of *myosin,* and the thin myofilaments have a more complex structure containing several proteins (*i.e.*, actin, tropomyosin, and troponins) of which *actin* is the most important. The two types of filaments are disposed in register and overlap to an extent that depends on the degree of contraction of the sarcomere. In a relaxed condition, the I-band contains only thin filaments; the H-band contain only thick filaments; and within the A-band the thick and thin filaments overlap.

In a cross section through the A-band, the regular disposition of the two types of filaments can be observed best. In vertebrate muscle each thick filament is seen to be surrounded by six thin filaments, and each thin filament lies symmetrically among three thick ones. As a consequence of this geometry there are twice as many thin filaments as thick ones. A cross section through the H-band shows only thick myofilaments and through the I-band, only thin myofilaments. Another interesting detail revealed by the electron microscope is that the two sets of filaments are linked together by a system of cross-bridges. These arise from the thick filaments at intervals of about 7 nm.

Each bridge is situated along the axis with an angular difference of 60 degrees. This means that the bridges describe a helix about every 43 nm. As a result of this arrangement one thick filament joins the six adjacent thin ones every 43 nm. A repeat period corresponding to this distance can be observed in the A-band after staining with uranyl acetate. There are 11 strips in each half an A-band starting from the M-line. Some of these are due to the myosin cross-bridges, but others, to the presence of C-protein.

The complex fine structure of the A-band can also be revealed by negative staining of ultrathin sections of frozen muscle (*i.e.*, cryosections). In cross section the M-line shows that the thick myofilaments are joined, forming a triangular lattice A-3.

The Z-line

In cross section the Z-disc shows a woven-basket lattice which remains essentially unchanged during contraction. This lattice is made of Z-filaments which are connected to the thin filaments of the I-band. It is presumed that as one thin filament enters the Z-line, it is in continuity with three curved Z-filaments which unite it with three other thin filaments of the same sarcomere. According to this model the thin filaments of the opposite sarcomere are similarly arranged.

The hexagonal lattice that is characteristic of the A-band (one thick filament surrounded by six thin ones) becomes compressed at the I-Z junction and is transformed into the square lattice characteristic of the woven-basket. This model, in which there is no interlooping of filaments from one sarcomere to the other, explains the splitting of the Z-line that may occur under certain conditions. Later one we shall see that one of the main components of the Z-line is a-*actinin*, a protein that joins with the actin thin filament and may constitute the Z-filaments.

Sacromere I-band

In a living muscle fibre, changes with contraction can be observed with the phase contrast and interference microscope. If this is done, one striking observation that can be made is that the A-band remains constant in a wide range of muscle lengths, whereas the I-band shortens in accordance with the contraction.

The shortening of the I-band is the result of the fact that the thin myofilaments slide farther and farther into the arrays of thick filaments. With the progressing contraction, the thin filaments penetrate into the H-band and may even overlap, thereby producing a more dense band in the centre of the sarcomere (*inversion of the banding*). Finally, the thick filaments make contact and are crumpled against the Z-lines. These findings have been interpreted in the so-called *sliding filament mechanism* of contraction, which will be discussed in greater detail later. The degree of contraction thus achieved can be measured by determining the length of the sarcomere (*i.e.*, the distance between Z-discs) at rest and when it has shortened. Note that insect muscle, in general, shortens only slightly (about 12 per cent), whereas the shortening in vertebrate muscle may be much greater (about 43 per cent).

Macromolecular Organisation

The electron microscope has revealed that "smooth" muscles may have a varied macromolecular organization. In many cases they contain thin and thick myofilaments, as do striated muscles, but the difference lies in the absence of the Z-line and the lack of periodicity. In mollusks and annelids there are muscles with a helical arrangement that have thin and thick myofilaments linked by cross bridges. In the adductor muscle of the oyster, the so-called paramyosin muscle, each thick filament is surrounded by 12 thin filaments. The smooth muscle of vertebrates apparently lacks these two types of myofilaments and even myofibrils are difficult to recognize. However, improvements in the preparative techniques have demonstrated coarse myofilaments. Furthermore, thick and thin myofilaments have been isolated. In smooth muscles the contraction is very slow, but extreme degrees of shortening may be achieved.

Contractive System

In recent years knowledge about the molecular machinery involved in muscle contraction has become increasingly complex. To the classic force-generation proteins *myosin* and *actin* others having a regulatory function have been added, such as *tropomyosin* and the various *troponins*.

Finally a series of other proteins, probably playing a structural role, such as a-*actinin* present in the Z-band, the *M-band proteins*, and the *C-protein*, have been described. Here, we shall only mention some of the main characteristics of these proteins in relation to the structure of the myofilaments and myofibrils.

Myosin Molecules

Myosin can be extracted from muscles with a 0.3 M solution of KCl and purified by precipitation at lower ionic strength. This large molecule comprises about half of the total protein of the myofibril and has a molecular weight of about 500,000 daltons. It contains 2 polypeptide chains of about 200,000 daltons each and 4 smaller ones in the range of 20,000 daltons.

The 2 heavy chains are coiled around each other, forming a double helical a-helix about 140 nm long and 2 nm in diameter. Only about half the heavy chain is rod-like, the rest, together with some of the smaller chains, is folded into two globular regions at one end cf the molecule. Myosin can be fragmented by the action of proteolytic enzymes; *trypsin* breaks it into the long, rod-like *light meromyosin* (LMM) and *heavy meromyosin* (HMM).

This last portion can be further subdivided by *papain* into the globular *subunit* S_1 and the *helical rod* S_2, which joins S_1 to the to the light meromyosin. The most important part of the myosin molecule is HMM-S_1, since it contains the sites for the *ATP-ase* and for *binding to actin*. In fact it is the interaction between myosin and actin that results in contraction.

At this point it is important to mention how the myosin molecules are integrated to constitute the *thick* or *myosin myofilament*. Each of these filaments has a smooth or bare portion in the middle (corresponding to the H-band) ad two regions in which the surface is rough because of the presence of the *myosin crossbridges*. It is now known that these *cross-bridges*. It is now known that these cross-bridges represent the S_1 globular ends of myosin, while the shaft of the filament is formed by the rod-like portion of the molecule.

One half of the molecules in the myofilament are oriented in the opposite direction from the other half. In other words, there is a definite polarization of the S_1 ends in each half of the myofilament. From the study of cross sections of the myosin filaments, which show a triangular shape, a model in which 12 parallel units are closely packed has been postulated. Each one of the units, with a diameter of 4 nm, could be the result of the association of two myosin molecules.

Another interesting observation is that at the M-line a cross section of the myosin filament shows bridges which connect with the six neighboring thick filaments, forming a hexagonal lattice. Such bridges represent the attachment of the M-protein at the M-line of the sarcomere.

Myofilament

Actin needs high concentrations of KCl (0.6 M) to be extracted. It is the second major protein and represents about one quarter of the protein of the myofibril. The molecular weight of actin is 42,000 daltons. The actin myofilament is made up of two helical strands which cross over every 36 to 37 nm.

Each of these cross-over repeats contains between 13 and 14 *globular* monomers of *G-actin* about 5 nm in diameter. In muscle, actin is present mainly as fibrous or *F-actin,* which is the polymerized form of G-actin. The actin myofilament is mre complex because it also contains *tropomyosin* and the *troponins.*

Tropomyosin

This protein represents between 5 and 10 per cent of the total and is extracted with 1 M KCI or weak acids. It has a molecular weight of 64,000 daltons and is an elongated molecule of 40 nm made of 2 parallel polypeptide chains in an a-helix. Tropomyosin stretches over seven actin monomers and forms two helixes that lie within the grooves of the actin double helix.

Troponin

Troponin is known to consist of 3 components forming a complex of proteins of 80,000 daltons. Each of these troponin complexes binds to tropomyosin every 40 nm, or every 7 actin monomers. The three components of troponin that are present in equimolecular proportions are *troponin-C,* which binds Ca^{2+} specifically, *troponin-T,* which has a high affinity for tropomyosin, and *troponin*-I, which inhibits the ATPase of myosin.

Contractile and Regulatory Proteins

Antibodies against myosin, actin, tropomyosin, and troponin that have been labeled with fluorescent dyes have permitted the localization of these proteins in relation to the macromolecular structure of the myofibril and myofilaments. For example, with *anti-troponin,* the fluorescent labeling of the thin filaments appears at a 40 nm interval. Other minor muscle proteins may also be localized with fluorescent antibodies.

A-Actinin

This protein is a rod-shaped molecule of about 95,000 daltons localized exclusively at the Z-line, as can be demonstrated by anti-actinin antibodies. The a-actinin molecule binds in vitro to actin filaments, forming regular cross-links. This protein apparently plays a structural role in the sarcomere.

C-Protein

This protein is present in the A-band along the middle portion of the myosin filament. With electron microscopy the anti-C-protein antibody is observed to form between 7 and 9 transverse stripes in each half of the A-band.

M-Line Protein

Several proteins have been extracted from the M-line region; one of them is probably a structural protein, and the others are phosphorylase and creatine-kinase.

Slioing Mechanism

We mentioned earlier that contraction is currently explained by a *sliding mechanism* in which the thin actin filaments are displaced with respect to the thick myosin filaments during each contraction-relaxation cycle. In the case of a frog sarcomere, 2.5 μm ong, there is a shortening of 30 per cent at each contraction. This implies a sliding of 0.37 μm for each half sarcomere. The force generated seems to be proportional to the degree of overlap between thick and thin filaments. This implies that the force is of a short range and is produced directly by the cross-bridges between the filaments. A resting muscle is rather plastic and extensible because the cross-bridges are not attached and can slide upon an applied external force.

Active sliding movement is thought to result from the repetitive interaction of the cross bridges with the actin filaments. It is assumed that each cross-bridge represents the active end of the myosin molecule (*i.e.*, the S_1 globular unit).

The following mechanism takes place:

- A cross-bridge binds to a specific site of the actin filament.
- The cross-bridge undergoes a conformational change which displaces the point of attachment towards the centre of the A-band, thereby pulling the acting filament; at the same time a second cross-bridge becomes attached.
- At the end of the cycle the first bridge returns to the starting configuration in preparation for a new cycle. According to this theory, the actin filament from each half sarcomere are pulled as ropes towards the centre of the sarcomere by the myosin arm that move to and from.

The energy for this interaction is provided by the splitting of ATP because of the ATPase present in the cross-bridge. It is estimated that the splitting of one ATP accounts for a displacement of 5 to 10 nm for each myosin cross-bridge.

Polarization within the Sarcomere

One of the most characteristic features of muscle is a molecular organization by which individual molecules may interact both spatially and temporally in a concerted fashion. Thus, the sliding mechanism depends on a very specifi interaction between actin and myosin molecules and requires the existence of a *definite polarization* within the sarcomere.

In each half sarcomere actin and myosin should have a different polarization; furthermore, the force developed must have opposite directions. From electron microscopic studies of aggregates of myosin molecules, it is possible to obtain reconstituted myofilaments having the structure. The molecules of each half of the filament are in two antiparallel sets with a reversal of structural polarity in the centre.

Muscle Contraction

Since the classic experiments of Szent-Györgyi, it has been known that the mixing of myosin and actin in the test tube results in the formation of the complex *actomyosin*, which contracts in the presence of ATP. Investigators following this interaction under the electron microscope have observed that the myosin molecules bind to the F-actin with a directional orientation. The complex can also be produced by using F-actin and heavy meromyosin.

The actin filaments become "decorated" with the myosin, and the complex shows an arrow-like polarity. These findings demonstrate that in each filament of F-actin, the G-actin molecules are polarized with the same orientation. The double helical filament of actin decorated with the S_1 myosin fragments. The arrow-like configuration shown in the electron micrograph results in the tilting and bending of the S_1 myosin along the helix of F_1 actin. This tilting of the S_1 myosin very likely corresponds to the configuration that the cross-bridge has at the end of the working stroke, when ADP and Pi have been released and no further force is being exerted.

On the other hand, there is electron microscopic and x-ray diffraction evidence that in the relaxed muscle the bridges are approximately perpendicular to the axis of the myofilaments. The detailed ultrastructural and biochemical information obtained permits an interpretation of the macromolecular mechanisms involved in muscle contraction. It may be postulated that this is a cyclic event involving the repetitive formation and breakdown of actin-myosin linkages at the bridges between thick and thin filaments.

At each bridge the following sequence of events is probably produced:

- Formation of a perpendicular linkage between a heavy meromyosin head and one G-actin (globular unit;
- Rupture of this linkage by one ATP molecule;
- Hydrolysis of the ATP by the Ca^{2+}-activated ATPase of myosin;
- Formation of a new linkage between the same heavy meromyosin (bridge) and the next G-actin unit.

The relative movement of the thin filament taking place in each sequence would be equivalent to the length of one G-actin unit (5.3 nm).

Energetics of Contraction

We have seen earlier that in the *relaxed state* there is no attachment of the cross-bridges; the opposite condition could explain the state of *rigor* (*i.e.*, of permanent contraction). In this state, the rigidity of the muscle could be due to the permanent attachment of the cross-bridges to the thin filaments. In order to have a contraction-relaxation cycle, a cyclic mechanism by which the cross-bridges become attached and detached must operate. It is now known that the transition between rest and activity is dependent on the *concentration of free calcium* in the vicinity of the contractile machinery.

The control of contraction by Ca^{2+} requires the presence of the regulatory proteins tropomyosin and troponin, which form part of the structure of the thin filament. We have seen earlier that *troponin-C* (Ca^{2+} binding) and *troponin-I* and *troponin-T* are placed along the actin filament at intervals of about 40 nm and that such a periodicity corresponds to 7 G-actin monomers. Furthermore, we mentioned that tropomyosin also extends along the actin filament for the same distance and is in close relationship with the troponin complex.

The mechanism by which Ca^{2+} regulates the contraction-relaxation cycle is thought to be the following:

- In the absence of Ca^{2+} the regulatory proteins inhibit the interaction between actin and myosin.
- When the Ca^{2+} concentration in the cytosol increases above 10^{-6} M, the muscle contraction is triggered by the binding of Ca^{2+} to troponin-C.
- Since the presence of tropomyosin is needed for this regulatory mechanism, it is supposed that the influence of troponin is transmitted by way of tropomyosin to the seven G-actin monomers with which it is associated. Thus, this mechanism is highly cooperative.

Molecular Regulation

From electron microscopic and x-ray diffraction evidence a model has been constructed which may explain the regulatory action of the tropomyosin-troponin-Ca^{2+} system. In this model, an end-on view of the relative positions of action, the S_1-myosin, and tropomyosin is presented. The tropomyosin molecule is shown in two positions, one corresponding to the activated state and the other to the relaxed state. According to the model in the latter condition (dotted contour) tropomyosin is deep in the groove of F-actin and covers the actin binding site of the S_1-myosin.

When Ca^{2+} binds to troponin-C displacement of tropomyosin away from the groove could be effected, exposing the actin binding site in S_1 for actin-myosin interaction.

Oxidative Phosphorylation

Whereas the myofibrils constitute the mechanical machinery of the muscle, the fuel needed is produced mainly in the sarcoplasm. In all types of muscle, numerous mitochondria, called *sarcosomes*, provide the essential oxidative phosphorylation processes and the Krebs cycle system. These mitochondria are particularly prominent in size and number in heart muscle and in the flight muscles of birds and insects.

Excitation and Contraction Coupling

Excitation-contraction coupling refers to the mechanism by which the electrical impulse is able to induce contraction in a muscle. It is known that with the arrival of the nerve impulse at the motor end plate a *chemical synaptic transmission* occurs. This in turn brings about a depolarization of the cell membrane which is conducted along the muscle fibre and penetrates into it to induce the contraction.

The sarcoplasmic matrix contains the glycolytic enzymes as well as other globular proteins, such as myoglobin, salts, and high phosphate compounds. Glycogen is present in the matrix as small granules observed under the electron microscope. There are about 1 per cent glycogen and 0.5 per cent creatine phosphate as sources of energy in muscle. Glycogen disappears with contraction through glycolysis, and lactic acid is formed, which can be transformed into pyruvic acid to enter the Krebs cycle. The initial energy source for contraction is ATP. The ADP produced after the initial contraction is again recharged to ATP by glycolysis or from creatine phosphate. Oxidative phosphorylation is the last and most important source of ATP.

You will see that the essential components in this excitation-contraction coupling mechanism are:

- The *T-system*, which conducts the action potential to The interior of the muscle fibre;

- The release of Ca^{2+} from the *sarcoplasmic reticulum*;
- The induction of contraction by Ca^{2+} (explained earlier);
- The reaccumulation of Ca^{2+} into the sarcoplasmic reticulum by Ca^{2+} -*ATPase*, which results in *muscle relaxation*. The study of this mechanism involves the knowledge of each one of the elements involved.

Sarcoplasmic Reticulum

The sarcoplasmic reticulum found in skeletal and cardiac muscle fibres is one of the most interesting specializations of the endoplasmic reticulum. It was discovered by Veratti in 1902 as a reticulum present in the sarcoplasm of the muscle fibre and extending in between the myofibrils. It was completely neglected until 1953 when the first electron micrographs of this structure were published.

The sarcoplasmic reticulum is a membrane-limited reticular system whose organizational structure is regularly superimposed upon that of the sarcomeres.

There are two main components in the sarcoplasmic reticulum:

1. A *longitudinal* component that is larger and represents the true endoplasmic reticulum;

2. A *transverse* component or *T-system* made of vesicles and tubules that are disposed at the level of the Z-lines and are connected to the sarcolemma.

The longitudinal reticulum is composed of wide anastomosing tubules that cover the surface of the sarcomere at the level of the A-and I-bands and that terminate by special terminal cisternae found at the end of the I-bands near the Z-line. In between the terminal cisternae of two consecutive sarcomeres is another flattened vesicle belonging to the T-system; the three together constitute the so-called *triad*. The T-system is in Continuity with the Plasma Membrane and Conducts Impulses Inward The transverse component (*i.e.*, the T-system at certain points is continuous with the plasma membrane of the sarcolemma and is the structure organized to conduct impulses from the fibre surface into the deepest portions of the muscle fibre. The continuity between the T-system, the plasma membrane, and the extracellular space was first demonstrated indirectly.

In frog muscle immersed for short periods in a solution containing ferritin molecules, it was found that the central vesicle of the triad had filled with these electron-opaque molecules. Ferritin was also found in certain tubules that were continuous with the central element of the triad.

These findings suggested that at the plasma membrane there are a small number of openings communicating directly with the traverse system of the sarcoplasmic reticulum through fine tubules. Interestingly, the later components of the triad never contained ferritin molecules, these components, then, not connected to the plasma membrane.

Direct observation of the connections between the T-system and the plasma membrane is difficult but has been achieved. Portion of the T-system have been observed to longitudinally and even in a spiral arrangement. The function of the T-system is to transverse inward the electrical signal that will bring about the contraction of the individual embroils. The role of the system was suggested experiments with microelectrodes in which stimulation of the sarcolemma, at the level the Z-line, produced a localized contraction adjacent sarcomeres.

Apparently the system is not a passive channel but is active involved in the conduction of action potential One interesting finding has been that in muscle immersed in glycerol there is a lesion of the T-system (*i.e., detubulation*), and the mechanism of excitation-coupling is abolished. The presence of this intracellular conducting system may explain the physiologic paradox that all the myofibrils of a fibre between 5 and 100 μm in diameter may contract quickly and synchronously, once the activating action potential has passed over the surface.

Stored Ca²⁺

Within the muscle, Ca^{2+} is stored mainly in the longitudinal components of the sarcoplasmic reticulum, so that the Ca^{2+} concentration near the myofibrils is kept very low (*i.e.*, about 10^{-7} M). Following stimulation, Ca^{2+} is rapidly released, so that the concentration increased (up to 10^{-5} M) and induces contraction. It is now assumed that the triad plays an essential role in this release of Ca^{2+}. It is thought that the action potential, arriving by way of the T-system, immediately opens Ca^{2+} channels in the *terminal cisternae* of the sarcoplasmic reticulum, producing a localized release of Ca^{2+}. Thus, the triad is the most important site in the excitation-contraction coupling.

We have read that during excitation Ca^{2+} is liberated from special regions of the sarcoplasmic reticulum and that this causes the contraction. To produce relaxation the opposite phenomenon must take place. In other words, Ca^{2+} must be rapidly removed, and this is done by a Ca^{2+} -*activated ATPase* that is present in the membrane of the sarcoplasmic reticulum. This acts as a Ca^{2+} pump that incorporates this cation under the action of ATP.

The reactions involved are the following:

$$2\ Ca^{2+}\ outside\ +\ ATP\ +\ E$$

$$E\text{-}PCa2^{2+}\ +\ ADP$$

$$E\text{-}PCa^{2+}\ E\text{-}Pi\ +\ 2Ca^{2+}\ inside$$

The two equations show that the splitting of one molecule of ATP results in the translocation of two Ca^{2+} by the ATPase (*E*). This reaction is reversible (*i.e.*, Ca^{2+} inside can be used for the formation of ATP).

The sarcoplasmic reticulum can be isolated by cell fractionation and vesicular structures may be obtained. The vesicles contain 40 per cent lipid, most of which is in the form of phosphatidylcholine, and two main proteins: the Ca^{2+} *pump protein* (*ATPase*), which, with a molecular weight of 105,000 daltons, represents about 80 per cent of the protein of the membrane, and a Ca^{2+} -*binding protein* (*i.e.*, *calsequestrin*), with a molecular weight of 65,000 daltons. The Ca^{2+} ATPase is thought to be a highly asymmetric molecule spanning the membrane, which needs the presence of a certain amount of phospholipid to function.

Since both the ATPase and the Ca^{2+} uptake are inhibited by agents that block –SH groups, an electron microscopic study was conducted using a cytochemical reagent in which ferritin molecules were attached to an –SH blocking molecule. It was demonstrated that the active groups of the membrane-bound ATPase were localized in the outer surface of the isolated sarcoplasmic reticulum. All these findings suggest that the sarcoplasmic reticulum assumes the important role of relaxing the fibre after contraction and that this is accomplished by the binding of Ca^{2+} at the outer surface and its transport within the sarcoplasmic vesicles. *Calsequestrin*, on the other hand, could be responsible for the binding of Ca^{2+} within the sarcoplasmic reticulum.

In summary, the series of events produced after the arrival of the electrical signal is as follows:

- The signal is received at the level of the Z-line or the A-I junction by way of the T-system.
- The coupling of the T-system with the terminal cisternae produces the release of Ca^{2+} (this may occur in a matter of milliseconds).
- Ca^{2+} induces the contraction, affecting the regulatory proteins troponin and tropomyosin and thus enabling the interaction of actin and myosin. During this step ATP is used as an energy source.
- Ca^{2+} is quickly removed and restored into the sarcoplasmic reticulum by way of the Ca^{2+} pump. Ca^{2+} uptake results in muscle relaxation.

All these data, as well as those related to the sliding mechanism of contraction, can be put together in a molecular theory of muscular contraction. This is one of the best examples, so far studied, of a tight coupling between the processes furnishing energy and the actual machinery involved in contraction. In this case, structure and function are so intimately related in the realm of molecular organization that they are an inseparable unit.

Cytogenetices

Cytogenetics is a branch of genetics that is concerned with the study of the structure and function of the cell, especially the chromosomes. It includes routine analysis of G-Banded chromosomes, other cytogenetic banding techniques, as well as molecular cytogenetics such as fluorescent in situ hybridization (FISH) and comparative genomic hybridization (CGH). During the last decade great advances have been made in the study of human chromosomes in the areas of their identification in the karyotype and the localization of genes.

Such studies have had important biological and medical implications in light of the discovery that many congenital disease and syndromes are related to chromosomal aberrations Today, *human cytogenetics* has becomes a specialized science in itself, the wide-ranging interests of which far exceed the limits of text on cell biology. Here, we shall present to the student an elementary account which could be supplemented by reading some of the general references.

Karyotype

Tissue cultures of fibroblasts, bone marrow, skin, and peripheral blood combined with the action of colchicine and hypotonic solutions to block mitosis at metaphases and to separate the chromosomes were used to study the human karyotype. An important technical advance has been the introduction of *phytohemagglutinin, which* induces lymphocytes to transform into lymphoblast-like cells that start to divide 48 to 72 hours after exposure. The strong mitogenic properties of this substance allowed the development of microtechniques which employ small amounts of blood Spreading the cells on a slide causes them to burst and to display all the chromosome which are usually studied in metaphase. Recently, higher resolution of the chromosome structure has been achieved by using banding techniques to study mitosis in prophase and prometaphase.

A *karotype* human metaphase chromosomes is usually obtained from microphotographs. The individual chromo-somes are cut out and there lined up by size with their respective partners. The technique can be improved by determining the so-called *centromeric index,* which is the ratio of the lengths of the long and short areas of the chromosome.

More recently, a system has been introduced that involves a computer-controlled microscope and several accessories that permit:

- Scanning of slides,
- Location of cells in metaphase,
- Counting chromosomes,
- Transmission of digitally expressed images for computation and storage.

All these steps, which can be carried out automatically, may help in making karyotypes more rapidly and in determining chromosomal aberrations. The 23 pairs are disposed in 7 groups (A through G) decreasing is size and having the characteristics described. For example, group A includes pairs 1 to 3, large almost metacentric chromosomes; group B comprises pairs 4 and 5, large submetacentric chromosomes, and so forth.

The X chromosome is in group C (pairs 6 to 12), medium-sized submitochondrial chromosomes; on the other hand, chromosome Y is in group G, together with pairs 21 and 22, small acrocentric chromosome. It is important to remember that pairs 21 and 22 satellites that correspond to nuclear organizers, while the Y chromsome lacks satellites. Furthermore group D (pairs 13 to 15) also contains acrocentric chromosomes with satellites that correspond to nuclear organizers.

Chapter 3

Genetic Effects of Cross-fertilization

There are several mechanisms promoting cross-pollination and, consequently, cross-fertilization.

The most important ones are:

- Dioecy, *i.e.* male and female gametes are produced by different plants.

  - Asparagus *Asparagus officinalis* L.
  - Spinach *Spinacia oleracea* L.
  - Papaya *Carica papaya* L.
  - Pistachio *Pistacia vera* L.
  - Date palm *Phoenix dactylifera* L.

- Monoecy, *i.e.* male and female gametes are produced by separate flowers occurring on the same plant.

  - Banana *Musa* spp.
  - Oil palm *Elaeis guineensis* Jacq.
  - Fig *Ficus carica* L.
  - Coconut *Cocos nucifera* L.
  - Maize *Zea mays* L.
  - Cucumber *Cucumis sativus* L.

  In musk melon (*Cucumis melo* L.) most varieties show andromonoecy, *i.e.* the plants produce both staminate flowers and bisexual flowers, whereas other varieties are monoecious.

- Protandry, *i.e.* the pollen is released before receptiveness of the stigmata.

  - Leek *Allium porrum* L.
  - Onion *Allium cepa* L.

- Carrot *Daucus carota* L.
- Sisal *Agave sisalana* Perr.
• Protogyny, *i.e.* the stigmata are receptive before the pollen is released.
  - Tea *Camellia sinensis* (L.) O. Kuntze
  - Avocado *Persea americana* Miller
  - Walnut *Juglans nigra* L.
  - Pearl millet *Pennisetum typhoides* L. C. Rich.
• Self-incompatibility, *i.e.* a physiological barrier preventing normal pollen grains fertilizing eggs produced by the same plant.
  - Cacao *Theobroma cacao* L.
  - Citrus *Citrus* spp.
  - Tea *Camellia sinensis* L. O. Kuntze
  - Robusta coffee *Coffea canephora* Pierre ex Froehner
  - Sugar beets *Beta vulgaris* L.
  - Cabbage, kale *Brassica oleracea* spp.
  - Rye *Secale cereale* L.
  - Many grass species, *e.g.* perennial ryegrass (*Lolium perenne* L.)
• Flower morphology
  - Fig *Ficus carica* L.
  - Primrose *Primula veris* L.
  - Common buckwheat *Fagopyrum esculentum* Moench.

and probably in the Bird of Paradise flower *Strelitzia reginae* Banks Effects with regard to the haplotypic and genotypic composition of a population due to (continued) reproduction by means of panmixis will now be derived for a so-called panmictic population.

Panmictic reproduction occurs if each of the next five conditions apply:
• Random mating
• Absence of random variation of allele frequencies
• Absence of selection
• Absence of mutation

- Absence of immigration of plants or pollen In the remainder of this section the first two features of panmixis are more closely considered.

Random Mating

Random mating is defined as follows: in the case of random mating the fusion of gametes, produced by the population as a whole, is at random with regard to the considered trait. It does not matter whether the mating occurs by means of crosses between pairs of plants combined at random, or by means of open pollination.

Open pollination in a population of a cross-fertilizing (allogamous) crop may imply random mating. This depends on the trait being considered. One should thus be careful when considering the mating system. This is illustrated in Example.

Example: Two types of rye plants can be distinguished with regard to their epidermis: plants with and plants without a waxy layer. It seems justifiable to assume random mating with regard to this trait. With regard to time of flowering, however, the assumption of random mating may be incorrect. Early flowering plants will predominantly mate *inter se* and hardly ever with late flowering plants. Likewise late flowering plants will tend to mate with late flowering plants and hardly ever with early flowering ones. With regard to this trait, so-called assortative mating occurs.

One should, however, realise that the ears of an individual rye plant are produced successively. The assortative mating with regard to flowering date may thus be far from perfect. Also, with regard to traits controlled by loci linked to the locus (or loci) controlling incompatibility, *e.g.* in rye or in meadow fescue (*Festuca pratensis*), perfect random mating will therefore probably not occur. Selection may interfere with the mating system. Plants that are resistant to an agent (*e.g.* disease or chemical) will mate *inter se* (because susceptible plants are eliminated). Then assortative mating occurs due to selection.

Crossing of neighbouring plants implies random mating if the plants reached their positions at random; crossing of contiguous inflorescences belonging to the same plant (geitonogamy) is, of course, a form of selfing.

Random mating does not exclude a fortuitous relationship of mating plants. Such relationships will occur more often with a smaller population size. If a population consists, generation after generation, of a small number of plants, it is inevitable that related plants will mate, even when the population is maintained by random mating.

Indeed, mating of related plants yields an increase in the frequency of homozygous plants, but in this situation the increase in the frequency of homozygous plants is also due to another cause: fixation occurs because of non-negligible random variation of allele frequencies. Both causes of the increase in homozygosity are due to the small population size (and not to the mode of reproduction).

This ambiguous situation, so far considered for a single population, occurs particularly when numerous small subpopulations form together a large superpopulation. In each subpopulation random mating, associated with non-negligible random variation of the allele frequencies, may occur, whereas in the superpopulation as a whole inbreeding occurs. Example provides an illustration.

Example. A large population of a self-fertilizing crop, *e.g.* an F2 or an F3 population, consists of numerous subpopulations each consisting of a single plant. Because the gametes fuse at random with regard to any trait, one may state that random mating occurs within each subpopulation.

At the level of the superpopulation, however, selfing occurs. Selfing is impossible in dioecious crops, *e.g.* spinach (*Spinacia oleracea*). Inbreeding by means of continued sister × brother crossing may then be applied. This full sib mating at the level of the superpopulation may imply random mating within subpopulations consisting of full sib families. Seen from the level of the superpopulation, inbreeding occurs if related plants mate preferentially. This may imply the presence of subpopulations, reproducing by means of random mating.

If very large, the superpopulation will retain all alleles. The increasing homozygosity rests on gene fixation in the subpopulations. If, however, only a single full sib family produces offspring by means of open pollination, implying crossing of related plants, then the population as a whole (in this case just a single full sib family) is still said to be maintained by random mating.

Absence of Random Variation of Allele Frequencies

The second characteristic of panmixis is absence of random variation of allele frequencies from one generation to the next. This requires an infinite effective size of the population, originating from an infinitely large sample of gametes produced by the present generation. Panmixis thus implies a deterministic model. In populations consisting of a limited number of plants, the allele frequencies vary randomly from one generation to the next. Models describing such populations are stochastic models.

Diploid Chromosome Behaviour and Panmixis

One Locus with Two Alleles

The majority of situations considered in this book involve a locus represented by not more than two alleles.

This is certainly the case in diploid species in the following populations:

- Populations tracing back to a cross between two pure lines, say, a single cross
- Populations obtained by (repeated) backcrossing (if, indeed, both the donor and the recipient have a homozygous genotype)

It is possibly the case in populations tracing back to a three-way cross or a double cross. It is improbable in other populations, like populations of cross-fertilizing crops, populations tracing back to a complex cross, landraces, multiline varieties. To keep (polygenic) models simple, it will often be assumed that each of the considered loci is represented by only two alleles. Quite often this simplification will violate reality. The situation of multiple allelic loci is explicitly considered.

If the expression for the trait of interest is controlled by a locus with two alleles A and a (say locus A-a) then the probability distribution of the genotypes occurring in the considered population is often described by Probability

$$\frac{Genotype}{aa \quad Aa \quad AA}$$
$$f_0 \quad f_1 \quad f_2$$

One may represent the probability distribution (in this book mostly the term genotypic composition will be used) by the row vector (f_0, f_1, f_2).

The symbol f_j represents the probability that a random plant contains j A-alleles in its genotype for locus A-a, where j may be equal to 0, 1 or 2. It has become custom to use the word genotype frequency to indicate the probability of a certain genotype and for that reason the symbol f is used.

The plants of the described population produce gametes which have either haplotype a or haplotype A. (Throughout this book the term haplotype is used to indicate the genotype of a gamete.) The probability distribution of the haplotypes of the gametes produced by the population is described by,

$$\text{Probability} \quad \frac{Haplotype}{a \quad A}$$

$$g_0 \quad g_1$$

The symbol g_j represents the probability that a random gamete contains j Aalleles in its haplotype for locus A-a, where j may be equal to 0 or 1.

The row vector (g_0, g_1) describes, in a condensed way, the haplotypic composition of the gametes. The habit to use the symbol q instead g_0 and the symbol p instead of g_1 is followed in this book whenever a single locus is considered. The term allele frequency will be used to indicate the probability of the considered allele.

So far it has been assumed that the allele frequencies are known and hereafter the theory is further developed without considering the question of how one arrives at such knowledge. In fact allele frequencies are often unknown.

When one would like to estimate them one might do that in the following way. Assume that a random sample of N plants is comprised of the following numbers of plants of the various genotypes: Number of Plants.

$$\frac{Genotype}{aa \quad Aa \quad AA}$$

$$n_0 \quad n_1 \quad n_2$$

For any value for N the frequencies q and p of alleles a and A may then be estimated as,

$$q = \frac{2n_0 + n_1}{2N} \text{ and } p = \frac{n_1 + 2n_2}{2N}$$

Throughout the book the expressions 'the probability that a random plant has genotype Aa', or 'the probability of genotype Aa', or 'the frequency of genotype Aa' are used as equivalents. This applies likewise for the expressions 'the probability that a gamete has haplotype A', or 'the probability of A'. Fusion of a random female gamete with a random male gamete yields a genotype specified by

j, the number of A alleles in the genotype. (The number of a alleles in the genotype amounts – of course – to 2-j.) The probability that a plant with genotype aa results from the fusion is in fact equal to the probability of the event that j assumes the value 0.

The quantity j assumes thus a certain value (0 or 1 or 2) with a certain probability. This means that j is a random variable. The probability distribution for j, $i.e.$ for the genotype frequencies, is given by the binomial probability distribution:

$$P(j = j) = \binom{2}{j} p^j q^{2-j}$$

Fusion of two random gametes therefore yields:
- With probability $q2$ a plant with genotype aa
- With probability $2pq$ a plant with genotype Aa
- With probability $p2$ a plant with genotype AA

The probabilities for the multinomial probability distribution of plants with these genotypes may be represented in a condensed form by the row vector ($q2$, $2pq$, $p2$). This notation represents also the genotypic composition to be expected for the population obtained after panmixis in a population with gene frequencies (q, p). In the case of panmixis there is a direct relationship between the gene frequencies in a certain generation and the genotypic composition of the next generation.

Thus if the genotype frequencies f_0, f_1 and f_2 of a certain population are equal to, respectively, q_2, $2pq$ and p_2, the considered population has the so-called Hardy–Weinberg (genotypic) composition. The actual genotypic composition is then equal to the composition expected after panmixis. With continued panmixis, populations of later generations will continue to have the Hardy–Weinberg composition. Therefore such composition may be indicated as the Hardy–Weinberg equilibrium. The names of Hardy and Weinberg are associated with this genotypic composition, but it was in fact derived by Castle in 1903. With two alleles per locus the maximum frequency of plants with the Aa genotype in a population originating from panmixis,

$$\frac{1}{2} \text{ for } p = q = \frac{1}{2}.$$

This occurs in F_2 populations of self-fertilizing crops. The F_2 originates from selfing of individual plants of the F_1, but because each plant of the

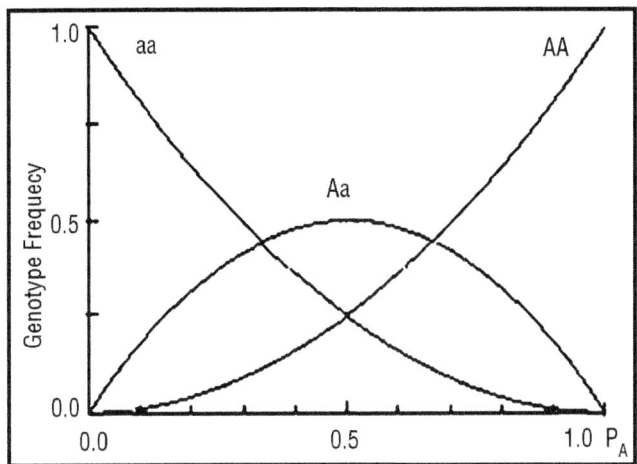

F_1 has the same genotype, panmixis within each plant coincides with panmixis of the F_1 as a whole. (The F1 itself may be due to bulk crossing of two pure lines; the proportion of heterozygous plants amounts then to 1.) The Hardy–Weinberg genotypic composition constitutes the basis for the development of population genetic theory for cross-fertilizing crops.

It is obtained by an infinitely large number of pairwise fusions of random eggs with random pollen, as well as by an infinitely large number of crosses involving pairs of random plants. One may also say that it is expected to occur both after pairwise fusions of random eggs and pollen, and when crossing plants at random.

In a number of situations two populations are crossed as bulks. One may call this bulk crossing. One population contributes the female gametes (containing the eggs) and the other population the male gametes (the pollen, containing generative nuclei in the pollen tubes). In such a case, crosses within each of the involved populations do not occur. A possibly unexpected case of bulk crossing is described in Note.

Note: Selection among plants after pollen distribution, *e.g.* selection with regard to the colour of the fruits (if fruit colour is maternally determined), implies a special form of bulk crossing: the rejected plants are then excluded as effective producers of eggs (these plants will not be harvested), whereas all plants (could) have been effective as producers of pollen. The results, to be derived hereafter, in the main text, for a bulk cross of two populations with different allele frequencies.

A bulk cross is of particular interest if the haplotypic composition of the eggs differs from the haplotypic composition of the pollen. Thus if population I, with allele frequencies (q_1, p_1), contributes the eggs and population II, with allele frequencies (q_2, p_2), the pollen, then the expected genotypic composition of the obtained hybrid population, in row vector notation, is

$$(q_1 q_2, p_1 q_2 + p_2 q_1, p_1 p_2)$$

This hybrid population does not result from panmixis. The frequency of allele A is,

$$P = \frac{1}{2}(p_1 q_2 + p_2 q_1) + p_1 p_2 = \frac{1}{2}p_1 q_2 + \frac{1}{2}p_1 p_2 + \frac{1}{2}p_2 q_1 + p_1 p_2$$

$$= \frac{1}{2}p_1(q_2 + p_2) + \frac{1}{2}p_2(q_1 + p_1) = \frac{1}{2}(p_1 + p_2)$$

$$q_2 + p_2 = q_1 + p_1 = 1$$

Further equations based on $p + q = 1$ are elaborated When deriving Equation the equation $p + q = 1$ was used. On the basis of the latter equation several other equations, applied throughout this book, can be derived:

$$q^2 + 2pq + p^2 = 1$$

$$p - q = 2p - 1 = 1 - 2q$$

$$(p - q)^2 = p^2 - 2pq + q^2) = 1 - 4pq$$

$$p^2 - q^2 = (p + q)(p - q) = p - q = f_2 - f_0$$

$$p - q + 2pq = p^2 - q^2 + 2pq = p^2 + 2pq - q^2 = 1 - 2q^2$$

Panmictic reproduction of this hybrid population produces offspring with the Hardy–Weinberg genotypic composition. The hybrid population contains, compared to the offspring population, an excess of heterozygous plants. The excess is calculated as the difference in the frequencies of heterozygous plants:

$$(p_1 q_2 + p_2 q_1) - 2pq = (p_1 q_2 + p_2 q_1) - 2[\frac{1}{2}(p_1 + p_2)\frac{1}{2}(q_1 + q_2)]$$

$$= \frac{1}{2}(p_1 q_2 + p_2 q_1 - p_1 q_1 - p_2 q_2)$$

$$= \frac{1}{2}(p_1 - p_2)(q_2 - q_1) = \frac{1}{2}(p_1 - p_2)^2$$

This square is positive, unless $p_1 = p_2$. Thus the hybrid does indeed contain an excess of heterozygous plants. Example illustrates that the superiority of hybrid varieties might (partly) be due to this excess. Example pays attention to the case of both inter- and intra-mating of two populations.

One Locus with more than Two Alleles

Multiple allelism does not occur in the populations considered so far. However, multiple allelism is known to occur in self- and cross-fertilizing crops. It may further be expected in three-way-cross hybrids, and their offspring, as well as in mixtures of pure lines (landraces or multiline varieties).

Example: The intensity of the anthocyanin colouration in lettuce (*Lactuca sativa*), a self-fertilizing crop, is controlled by at least three alleles. The colour and location of the white leaf spots of white clover (*Trifolium repens*), a cross-fertilizing crop, are controlled by a multiple allelic locus. The expression for these traits appears to be controlled by a locus with at least 11 alleles. Another locus, with at least four alleles, controls the red leaf spots. (White clover is an autotetraploid crop with a gametophytic incompatibility system and a diploid chromosome behaviour; $2n = 4x = 32$).

The frequencies (*f*) of the genotypes $AiAj$ (with $i \leq j$; $j = 1..., n$) for the multiple allelic locus $A1$-$A2$-... -An attain their equilibrium values following a single round of panmictic reproduction. The genotypic composition is then: *f*

Genotype		
A_1A_1...	$AjAj$...	A_nA_n
$p1^2$	$2pipj$	pn^2

The Proportion of homozygous plants is minimal for,

$$pj = \frac{1}{n}(for\ j = 1,...,n)$$

and amounts then to $n\left(\frac{1}{n}\right)^2 = \frac{1}{n}$;

Two Loci, Each with Two Alleles

Iit is shown that a single round of panmictic reproduction produces immediately the Hardy–Weinberg genotypic composition with regard to a single locus. It is immediately attained because the random fusion of pairs of gametes implies random fusion of separate alleles, whose frequencies are constant from one generation to the next. For complex genotypes, *i.e.* genotypes with regard to two or more loci (linked or not), however, the so-called linkage equilibrium is only attained after *continued panmixis*. Presence of the Hardy–Weinberg genotypic composition for separate loci does not imply presence of linkage equilibrium! (Example illustrates an important exception to this rule.)

In panmictic reproduction the frequencies of complex genotypes follow from the frequencies of the complex haplotypes. Linkage equilibrium is thus attained if the haplotype frequencies are constant from one generation to the next. For this reason 'linkage equilibrium' is also indicated as gametic phase equilibrium. In this section it is derived how the haplotypic frequencies approach their equilibrium values in the case of continued panmixis. This implies that the tighter the linkage the more generations are required. However, even for unlinked loci a number of rounds of panmictic reproduction are required to attain linkage equilibrium.

The genotypic composition in the equilibrium does not depend at all on the strength of the linkage of the loci involved. The designation 'linkage equilibrium' is thus not very appropriate.

To derive how the haplotype frequencies approach their equilibrium. We consider loci A-a and B-b, with frequencies p and q for alleles A and a and frequencies r and s for alleles B and b. The recombination value is represented by rc. This parameter represents the probability that a gamete has a recombinant haplotype. Independent segregation of the two loci occurs at $rc = 1/2$, absolute linkage at $rc = 0$. Example illustrates the estimation of rc in the case of a testcross with a line with a homozygous recessive (complex) genotype.

The haplotype frequencies are determined at the meiosis. The haplotypic composition of the gametes produced by generation Gt– 1 is described by

Haplotype			
ab	aB	Ab	AB
$g_{00,t}$	$g_{01,t}$	$g_{10,t}$	$g_{11,t}$

The last subscript (t) in the symbol for the haplotype frequencies indicates the rank of the generation to be formed in a series of generations generated by panmictic reproduction ($t = 1, 2...$).

Example. The spinach variety Wintra is susceptible to the fungus *Peronospora spinaciae* race 2 and tolerant to Cucumber virus 1. It was crossed with spinach variety Nores, which is resistant to *P. spinaciae* race 2 but sensitive to Cucumber virus 1.

The loci controlling the host-pathogen relations are *A-a* and *B-b*. The genotype of Wintra is *aaBB* and the genotype of Nores *AAbb*. The offspring, with genotype *AaBb*, were crossed with the spinach variety Eerste Oogst (genotype *aabb*), which is susceptible to *P. spinaciae* race 2 and sensitive to Cucumber virus 1. On the basis of the reaction to both pathogens a genotype was assigned to each of the 499 plants resulting from this testcros:

Genotype:

aabb	aaBb	Aabb	AaBb	Total
61	190	194	54	499
124.75	124.75	124.75	124.75	499

The expected frequencies are calculated on the basis of the null hypothesis stating that the two involved loci are unlinked. The expected 1 2: 1 2 segregation ratio was confirmed by a goodness of fit test for each separate locus. The specified null hypothesis is, of course, rejected. The two loci are clearly linked. The value estimated for rc is,

$$\frac{61+54}{499} = 0.23$$

Note: In this book the last subscript in the symbols for the genotype and haplotype frequencies indicate the generation number. If it is t it refers to population Gt, *i.e.* the population obtained by panmictic reproduction of t successive generations.

Population G1, resulting from panmictic reproduction in a single-cross hybrid, has the same genotypic composition as the F2 population resulting from selfing plants of the single-cross hybrid. To standardize the numbering of generations of cross-fertilizing crops and those of self-fertilizing crops, the population resulting from the first reproduction by means of selfing might be indicated by S1 (rather than by the more common indication F2). To avoid confusion this will only be done when appropriate. The last subscript in the symbols for the haplotype frequencies of the gametes giving rise to S1 are taken to be 1. The same applies to the frequencies of the genotypes in S1. This system for labelling generations of gametophytes and sporophytes was also adopted by Stam.

Population G0 is thus some initial population, obtained after a bulk cross or simply by mixing. It produces gametes with the haplotypic composition ($g00,1$; $g01,1$; $g10,1$; $g11,1$). In the absence of selection, allele frequencies do not change. This implies,

$$g_{10,1} + g_{11,1} = g_{10,2} + g_{11,2} = \ldots = p$$

for allele A, and similar equations for the frequencies of alleles a,B and b. It was already noted that the haplotype frequencies in successive generations will be considered. In the appendix of this section it is shown that the following recurrent relations apply:

$$g_{100,t} + 1 = g_{100,t} - r_c d_t$$
$$g_{01,t} + 1 = g_{01,t} + r_c d_t$$
$$g_{10,t} + 1 = g_{10,t} + r_c d_t$$
$$g_{11,t} + 1 = g_{11,t} - r_c d_t$$

Where the definition of d_t follows from:
$$2d_t := f_{11c,t} - f_{11R,t}$$

where ':=' means: 'is defined as', and t = 1, 2, 3...

In Note: it is shown that Equations also apply to selffertilizing crops. The recurrent equations show that the haplotype frequencies do not change from one generation to the next if $rc = 0$ or if $dt = 0$.

Such constancy of the haplotypic composition implies constancy of the genotypic composition. It implies presence of linkage equilibrium. Linkage equilibrium is thus immediately established by a single round of panmictic reproduction for loci with $rc = 0$. This situation coincides with the case of a single locus with four alleles.

The symbol $f11C$ indicates the frequency of AB/ab-plants, *i.e.* doubly heterozygous plants in coupling phase (C-phase); the symbol $f11R$ represents the frequency of Ab/aB-plants, *i.e.* doubly heterozygous plants in repulsion phase (R-phase).

In the case of panmixis the following equations apply:

$$f11c,t = 2(g11,tg00,t)$$
$$f11r,t = 2(g10,tg01,t)$$

In that case we get,

$$d_t = (g11,tg00,t) - (g10,tg01,t)$$

This parameter is called coefficient of link age disequilibrium. It Appears in the following derivation:

$$g_{11},t = g_{11},t(g_{10},t + g_{01},t + g_{11}, t + g_{00},t)$$
$$= (g_{10},tg_{01},t + g_{10},tg_{11},t + g_{11},tg_{01},t + g_{11},t)$$
$$+ (g_{11},tg_{00},t - g_{10},tg_{01},t)$$
$$= (g_{10},t + g_{11},t)(g_{01},t + g_{11},t) + d_t = p_r + d_t$$

Equation may thus be rewritten as

$$pr + d_{t+1} = (pr + d_t) - r_c d_t$$

Which implies not only

$$d_{t+1} = (1 - r_c)d_t$$

equation may thus be rewritten as

but of course also

$$d_t = (1 - r_c)^{t-1}d_1$$

for $t = 2,3,....$

The derivation above (and similar derivations for the

other haplotype frequencies)implies

$$d_t = g11,t - pr = -(g10,t - ps) = -(g01,t - qr) = g00,t - qs$$

Because $1/2 \leq (1-rc) \leq 1$, continued panmixis implies continued decrease of dt. The decrease is faster for smaller values of 1-rc, i.e. for higher values of rc. Independent segregation, i.e. $rc = 1/2$, yields the fastest reduction, viz. halving of dt by each panmictic reproduction. The value of dt eventually attained, i.e. $dt = 0$, implies that linkage equilibrium is attained, i.e. constancy of the haplotype frequencies. The haplotype frequencies have then a special value, viz.

$$g00 = qs$$
$$g01 = qr$$
$$g10 = ps$$
$$g11 = pr$$

The equilibrium frequencies of the haplotypes are equal to the products of the frequencies of the alleles involved, and the equilibrium frequencies of the complex genotypes are equal to the products of the Hardy–Weinberg frequencies of the single-locus genotypes for the loci involved. The strength of the linkage between the loci is irrelevant with regard to the genotypic composition in the equilibrium. It only affects the number of generations of panmictic reproduction required to 'attain' the equilibrium.

(a) *Genotypes*

	bb	Bb	BB	
aa	q^2s^2	$2q^2rs$	q^2r^2	q^2
Aa	$2pqs^2$	$4pqrs$	$2pqr^2$	$2pq$
AA	p^2s^2	$2p^2rs$	p^2r^2	p^2
	s^2	$2rs$	r^2	1

(b) *Phenotypes*

	bb	B	
aa	q^2s^2	$q^2(1-s^2)$	q^2
A	$(1-q^2)s^2$	$(1-q^2)(1-s^2)$	$(1-q^2)$
	s^2	$1-s^2$	

The foregoing is illustrated in Example, which deals with the production of a single-cross hybrid variety and the population resulting from its offspring as obtained by panmictic reproduction. Example illustrates the production of a synthetic variety and a few of its offspring generations as obtained by continued random mating.

GenotypicComposition	G_1 for G_0 in C - phase	G_1 for G_0 in R - Phase	$G\infty$
Genotype	$\frac{1}{4}(1-r_c)^2$	$\frac{1}{4}r_c^2$	$\frac{1}{16}$
aabb	$\frac{1}{2}r_c(1-r_c)$	$\frac{1}{2}r_c(1-r_c)$	$\frac{2}{16}$
aaBb	$\frac{1}{4}r_c^2$	$\frac{1}{4}(1-r_c)^2$	$\frac{1}{16}$
aaBB	$\frac{1}{2}r_c(1-r_c)$	$\frac{1}{2}r_c(1-r_c)$	$\frac{2}{16}$
Aabb	$\frac{1}{4}r_c^2$	$\frac{1}{2}r_c^2$	$\frac{2}{16}$
AB/ab	$\frac{1}{2}r_c(1-r_c)$	$\frac{1}{2}(1-r_c)^2$	$\frac{2}{16}$
Ab/aB	$\frac{1}{2}(1-r_c)^2$	$\frac{1}{2}rc(1-r_c)^2$	$\frac{2}{16}$
AaBB	$\frac{1}{4}r_c^2$	$\frac{1}{4}(1-r_c)^2$	$\frac{1}{16}$
AAbb	$\frac{1}{2}r_c(1-r_c)$	$\frac{1}{2}rc(1-r_c)$	$\frac{2}{16}$
AABb	$\frac{1}{4}(1-r_c)^2$	$\frac{1}{4}r_c^2$	$\frac{1}{16}$
AABB			

On the basis of the frequencies of the phenotypes for two traits (each with two levels of expression) showing qualitative variation, one can easily determine whether or not a certain population is in linkage equilibrium. It is, however, impossible to conclude whether or not the loci involved are linked.

Only test crosses between individual plants with the phenotype $A \cdot B \cdot$ and plants with genotype $aabb$ will give evidence about this. $N.B.$ By 'phenotype $A \cdot B$' is meant the phenotype due to genotype $AABB$, $AaBB$, $AABb$ or $AaBb$

Example: A synthetic variety is planned to be produced by intermating five clones of a self-incompatible grass species. Because crosses within each of the five components are excluded, the synthetic variety is produced by outbreeding. It is, therefore, due to a complex bulk cross. The obtained plant material is designated as Syn1 (or G0 in the present context). The five clones have the following genotypes for the two unlinked loci $B1$-$b1$ and $B2$-$b2$: clone 1: $b1b1b2b2$; clones 2 and 3: $B1B1b2b2$, and clones 4 and 5: $B1B1B2B2$.

The genotypic composition of Syn1 can be derived from the following scheme:

	$b_1b_1b_2b_2$	$B_1B_1b_2b_2$	$B_1B_1b_2b_2$	$B_1B_1B_2B_2$	$B_1B_1B_2B_2$
$b_1b_1b_2b_2$		$B_1b_1b_2b_2$	$B_1b_1b_2b_2$	$B_1b_1B_2b_2$	$B_1b_1B_2b_2$
$B_1B_1b_2b_2$	$B_1b_1b_2b_2$	-	$B_1B_1b_2b_2$	$B_1B_1B_2b_2$	$B_1B_1B_2b_2$
$B_1B_1b_2b_2$	$B_1b_1b_2b_2$	$B_1B_1b_2b_2$		$B_1B_1B_2b_2$	$B_1B_1B_2b_2$
$B_1B_1B_2B_2$	$B_1b_1B_2b_2$	$B_1B_1B_2b_2$	$B_1B_1B_2b_2$		$B_1B_1B_2B_2$
$B_1B_1B_2B_2$	$B_1b_1B_2b_2$	$B_1B_1B_2b_2$	$B_1B_1B_2b_2$	$B_1B_1B_2B_2$	

Table presents the genotype frequencies in a few relevant generations. When deriving these it was assumed that incompatibility can be neglected when considering continued panmictic reproduction starting in G0. The portion of homozygous plants in G0,G1,G2 and G– amounts to 0.2; 0.35; 0.3508 and 0.3536, respectively.

The excess of heterozygous plants in comparison to the linkage equilibrium amounts therefore to 0.1536; 0.0036 and 0.0028 in G0,G1 and G2, respectively. (This concerns plants which are heterozygous for one or two loci. For each single locus the Hardy–Weinberg genotypic composition occurs in G1 and all later generations). The genotypic composition of plant material obtained when creating and maintaining an imaginary synthetic variety. P indicates the parental clones, Go indicates population syn_1, G_1 indicates syn_2, G_2 indicates syn_3 and G– indicates syn–

Frequency

Genotype	P	G_0	G_1	G_2	$G\infty$
$b_1b_1b_2b_2$	0.2		0.0225	0.0182	0.0144
$b_1b_1B_2B_2$			0.0150	0.0176	0.0192
$b_1b_1B_2B_2$			0.0025	0.0042	0.0064
$B_1b_1b_2b_2$	0.2	0.1350	0.1256	0.1152
B_1B_2/b_1b_2	0.2	0.1050	0.0904	0.0768
B_1b_2/b_1B_2			0.0450	0.0605	0.0768
$B_1b_1B_2B_2$			0.0350	0.0436	0.0512
$B_1b_1b_2b_2$	0.4	0.1	0.2025	0.2162	0.2304
$B_1B_1B_2b_2$		0.4	0.3150	0.3116	0.3072
$B_1B_1B_2B_2$	0.4	0.1	0.1225	0.1122	0.1024

The Haplotype Frequencies in Generation t

In this appendix, first is derived an equation relating the frequency of gametes with haplotype ab in generation t + 1 to its frequency in generation t, *i.e.* Equation. Thereafter an equation describing the haplotype frequencies in generations due to continued panmictic reproduction, starting with a singlecross hybrid, is derived.

The Frequency of Gametes with Haplotype ab

The frequency of gametes with haplotype ab, produced by generation Gt, are equal to,

$$g00, t+1 = f00, t + \frac{1}{2}f10, t + \frac{1}{2}01, t + \frac{1}{2}(1-r_c)f11c, t + \frac{1}{2}r_c f11R, t$$

$$= f00, t + \frac{1}{2}f10, t + \frac{1}{2}f01, t + \frac{1}{2}f11c, t - r_c d_t$$

Panmictic reproduction of generation Gt yields generation $Gt + 1$. The genotypic composition of $Gt + 1$ is described by the frequencies given by the third column of the previous table. Inclusion of these genotype frequencies in the following equations: The relevant genotypes, their frequencies (in general, as well as after panmixis) and the haplotypic composition of the gametes they produce are:

Above equation for $g_{00,t+1}$ gives,

$$g00, t+1 = g^2 00, t + g00, tg10, t + g00, tg01, t + g00, tg11, t - r_c d_t$$

$$= goo, t(g00, t + g10, t + g01, t + g11, t) - r_c d_t = g00, t - r_c d_t$$

Where, according to Equation,

$$d_t = (g11, tg00, t - g10, tg01, t)$$

Genotype	in general	After Panmixis	ab	aB	Ab	AB
aabb	f00	$g00^2$	1	0	0	0
Aabb	f10	$2g00g10$	$\frac{1}{2}$	0	$\frac{1}{2}$	0
AAbb	f20	$g10^2$	0	0	1	0
aaBb	f01	$2g00g01$	$\frac{1}{2}$	$\frac{1}{2}$	0	0
$\dfrac{AB}{aB}$	f11C	$2g00g11$	$-\frac{1}{2}r_c$	$\frac{1}{2}r_c$	$\frac{1}{2}r_c$	$\frac{1}{2}$
$\dfrac{AB}{aB}$	f11R	$2g10g01$	$\frac{1}{2}r_c$	$\frac{1}{2}$	$\frac{1}{2}$	$-\frac{1}{2}r_c$
AABb	f21	$2g01g11$	0	$-\frac{1}{2}r_c$	$-\frac{1}{2}r_c$	$\frac{1}{2}r_c$
aaBB	f02	$g01^2$	0	0	$\frac{1}{2}$	$\frac{1}{2}$
AaBB	f12	$2g01g11$	0	1	0	0
AABB	f22	$g11^2$	0	$\frac{1}{2}$	0	$\frac{1}{2}$

Similarly one can derive,

$$g01,t+1 = g01,t + r_c d_t$$

$$g10,t+1 = g10,t + r_c d_t$$
$$g11,t+1 = g11,t - r_c d_t$$

The haplotype frequencies in generations due to continued panmictic reproduction, starting with a single-cross hybrid In the case of panmictic reproduction starting from a single-cross hybrid there will be a symmetry in the haplotype frequencies such that,

$$g00, t = g11, t$$

and,

$$g01, t = g10, t = \frac{1}{2} - g11, t$$

Derivation of $g11,t$ suffices then to obtain the frequencies of all haplotypes with regard to two segregating loci. An equation presenting $g11,t$ immediately for any value for t will now be derived. If the genotype of the single-cross hybrid is AB/ab, i.e. coupling phase, the genotypic composition of the initial population G0 is simply described by

- $f11C,0 = 1$, if it is Ab/aB the genotypic composition of G0 is described by $f11R,0 =$ Equation yields then

$$d_0 = \frac{1}{2}$$

in *the former case, and*

$$d_0 = \frac{-1}{2}$$

in the latter case. The frequency of gametes with the AB haplotype among the gametes produced by the single-cross amounts to,

$$g11,1 = \frac{1}{2}(1 - r_c) \text{ and } g11,1 = \frac{1}{2}r_c$$

Respectively in Example it was also derived that

$$d_1 = \frac{1}{4}(1 - 2r_c)$$

for G_0 in $C -$ Phase and that

$$d_1 = \frac{-1}{4(1 - 2r_c)}$$

respectively. In Example it was also derived that for G0 in R-phase. The frequencies of AB haplotypes in the case of continued panmixis follow from Equation combined with Equation:

$$g11, t+2 = g11, t+1 - r_c d_t + 1 = g11, t+1 - r_c(1 - r_c)^t d_1$$
$$= g11, t - rc(1 - rc)^{t-1}d_1 - r_c(1 - rc)^t d1$$
$$= g11,1 - r_c d_1[(1 - r_c)^0 + ... + (1 - r_c)^t]$$

The terms within the brackets form a convergent geometric series. The sum of such terms is given by the expression,

$$a\frac{1 - q^n}{1 - q}$$

Where a is the first term, q is the multiplying factor and n is the number of terms. In the present situation this sum amounts to,

$$\frac{1 - (1 - r_c)^{t+1}}{r_c}$$

Thus $g11, t+2 = g11,1 - d_1[1 - (1 - r_c)^{t+1}]$

For $r_c = \frac{1}{2}$ *we got $d_1 = 0$. Then*

$$g11, t+2 = g1,1 = \frac{1}{4}$$

This implies that linkage equilibrium is present after one generation with panmictic reproduction!

For G0 in C-phase, Equation can be rewritten as,

$$g11,t+2=\frac{1}{2}(1-r_c)-\frac{1}{4}(1-2r_c)[1-(1-r_c)^{t+1}]$$

Thus

$$g11,2=\frac{1}{2}(1-r_c)-\frac{1}{4}r_c(1-2r_c)=\frac{1}{2r}r_c^2-\frac{3}{4}r_c+\frac{1}{2}$$

For G_0 in R-Phase, Equation can be transformed into,

$$g11,t+2=\frac{1}{2}r_c+\frac{1}{4}(1-2r_c)[1-(1-r_c)^{t+1}]$$

This implies:

$$g11,2=\frac{1}{2}r_c+\frac{1}{4}r_c(1-2r_c)=-\frac{1}{2}r_c^2+\frac{3}{4}r_c$$

$$g11,3=\frac{1}{2}r_c+\frac{1}{4}(1-2r_c)[1-(1-r_c)^2]=\frac{1}{2}r_c^3-1\frac{1}{4}r_c^2+r_c$$

These equations are of relevance with regard to the question of whether it is advantageous, when it is aimed to promote the frequency of plants with a genotype due to recombination, to apply random mating in an F2 population of a self-fertilizing crop.

More than Two Loci, Each with Two or more Alleles

Linkage Involving Three Loci: Three loci A-a, B-b and C-c are considered. These loci occur in this order along a chromosome. The segments AB,BC and AC are distinguished. Effective recombination of alleles belonging to loci A-a and B-b requires that the number of crossover events in segment AB is an odd number. The probability of recombination is called recombination value, designated by the symbol rc, or by the symbol rAB or simply by r (depending on the context). With an even number of times of crossing-over in segment AB there is no (effective) recombination. The probability of this event is 1-rAB. There is (effective) recombination of alleles belonging to loci A-a and C-c if there is either (effective) crossing-over in segment AB, but not in segment BC; or if there is (effective) crossing-over in segment BC, but not in segment AB.

If the occurrence of recombination in one chromosome segment has no effect on the recombination value for the adjacent segment the following relation applies:

$$rAC=rAB(1-rBC(1-rAB)=rAB+rBC-2rABrBC$$

This situation is likely for loci that are not too closely linked. The situation where recombination in one segment depresses the probability of recombination in an adjacent segment is called chiasma interference.

A more general expression for rAC is thus:

$$rAC = rAB + rBC - 2(1 - \delta)rABrBC$$

where δ is the interference parameter, ranging from 0 (no interference) through 1 (complete interference). It shows that rAC is higher at higher values for δ. Recombination values are additive if

$$2(1 - \delta)rABrBC = 0$$

i.e. if $\delta = 1$ and/or $rABrBC = 0$. In other cases they are not additive. These conditions imply that recombination values are mostly not additive. They are, consequently, inappropriate to measure distances between loci. The hypothesis of independence of crossing-over in segments *AB* and *BC*, *i.e.* the hypothesis of absence of chiasma interference, can be tested by means of a goodness-of-fit test. Among N plants, the expected number of plants with a genotype which is due to double crossing-over amounts, according to this hypothesis, to $rABrBCN$. It is compared to the observed number.

The ratio,

$$\frac{Observed\ Number}{Expected\ Number}$$

is called coefficient of coincidence. When there is independency it is equal to 1. Its complement, *i.e.*

$$1 - \frac{Observed\ Number}{Expected\ Number}$$

Estimates δ. Its value is positive if the observed number of plants with the recombinant genotype is smaller than the number expected at independency: the presence of a chiasma in the one segment hinders the formation of a chiasma in the other segment. The actual distance between loci, say the map distance m, measures the total number of cross-over events (both odd and even numbers) between the loci. This distance is an additive measure. It can only approximately be determined from recombination values. Haldane (1919) developed an approximation for the situation in the absence of interference ($\delta = 0$). His mapping function is

$$m = -\frac{In(1 - 2r_c)}{2},$$

Where m represents the expected number of cross-over events in the considered segment. As the map distance is mostly expressed in centiMorgans (cM), this function is,

$$m = -50\ln(1 - 2r_c)$$

Often written as An approximation which takes interference into account is called Kosambi's mapping function.

Frequencies of Complex Genotypes after Continued Panmixis

It can be shown that continued panmixis eventually leads to an equilibrium of the frequencies of complex genotypes for three or more loci, each with two or more alleles. The equilibrium is characterized by haplotype frequencies equal to the products of the frequencies of the alleles involved. Linkage equilibrium for one or more pairs of loci does not imply equilibrium of the frequencies of complex genotypes for three or more loci. Equilibrium of the frequencies for complex genotypes implies, however, linkage equilibrium for all pairs of loci.

Autotetraploid Chromosome Behaviour and Panmixis

The implications of panmixis in an autotetraploid crop will only be considered for a single locus with two alleles. This is to keep the mathematical derivations simple. It will be shown that the equilibrium frequencies of the genotypes are not obtained after a single panmictic reproduction. At equilibrium the frequencies of the genotypes and the haplotypes are equal to the products of the frequencies of the alleles involved.

Among cross-fertilizing autotetraploid crops the more important representatives are alfalfa (*Medicago sativa* L.; $2n = 4x = 32$) and cocksfoot (*Dactylis glomerata* L.; $2n = 4x = 28$). Additionally, highbush blueberry (*Vaccinium corymbosum* L.; $2n = 4x = 48$) might be mentioned. Leek (*Allium porrum* L.; $2n = 4x = 32$) is an autotetraploid crop with a tendency to a diploid behaviour of the chromosomes. Among ornamentals several autotetraploid species occur, *e.g. Freesia hybrida, Cyclamen persicum* Mill. ($2n = 4x = 48$) and *Begonia semperflorens*. Also, artificial autotetraploid crops have been made, *e.g.* rye (*Secale cereale* L.; $2n = 4x = 28$) and perennial rye grass (*Lolium perenne* L.; $2n = 4x = 28$). In 1977 about 500,000 ha of autotetraploid rye were grown in the former Soviet Union. Sweet potato, *i.e. Ipomoea batatas* var. *littoralis* ($2n = 4x = 60$) or *I. batatas* var. *batatas* ($2n = 6x = 90$), may be considered as a cross-fertilizing crop (due to self-incompatibility), but it is mainly vegetatively propagated. Under certain conditions double reduction may occur in autotetraploid crops, in which case (parts of) sister chromatids end up in the same gamete. The resulting haplotype is homozygous for the loci involved. The process of double reduction causes the frequency of homozygous genotypes and haplotypes to be somewhat higher than in absence of double reduction.

Blakeslee, Belling and Farnham discovered the phenomenon in autotetraploid jimson weed (*Datura stramonium* L.; $2n = 4x = 48$): a triplex plant (with genotype *AAAa*) produced some nulliplex offspring after crossing with a nulliplex (genotype *aaaa*). This is only possible if the triplex plant produces *aa* gametes.

The process of double reduction is an interesting phenomenon, but in a quantitative sense it is of no importance. For this reason we assume that double reduction does not occur. The autotetraploid genotypes to be distinguished for locus *A-a* are *aaaa* (nulliplex), *Aaaa* (simplex), *AAaa* (duplex), *AAAa* (triplex) and *AAAA* (quadruplex).

In each cell these genotypes contain *JA* alleles and 4-*Ja* alleles. At meiosis two of these four alleles are sampled to produce a gamete.

The haplotypes that can be produced by an autotetraploid plant containing *JA* alleles can be described by *j*, the number of *A* alleles that they contain, where $j = 0$, 1 or 2. The conditional probability distribution for *j*, given that the parental genotype contains *JA* alleles, is a hypergeometric probability distribution:

$$P(\underline{j}=j/J)=\frac{\binom{J}{j}\binom{4-J}{2-j}}{\binom{4}{2}}=\frac{1}{6}\binom{J}{j}\binom{4-J}{2-j}$$

The Probability that a triplex plant (*i.e.* j =3) Produces a gemete with haplotype Aa (*i.e.* j = 1) is therefore,

$$P(\underline{j}=1/J=3)=\frac{1}{6}\binom{3}{1}\binom{1}{1}=\frac{1}{2}$$

Table Presents, for each autotetraploid genotype, the haplotypic composition, *i.e.* the probability distribution for the haplotypes produced. The genotypic composition of a tetraploid population is described like that of a diploid population. Thus in the case of autotetraploid species the row.

Vector (*f*0, *f*1, *f*2, *f*3, *f*4) is used. The equilibrium frequencies of the genotypes are attained as soon as the haplotype frequencies are stable. Therefore the haplotypic composition of successive generations with panmictic reproduction will be monitored.

Genotype	aa	Aa	AA
aaaa	1	0	0
Aaaa	$\frac{1}{2}$	$\frac{1}{2}$	0
AAaa	$\frac{1}{6}$	$\frac{4}{6}$	$\frac{1}{6}$
AAAa	0	$\frac{1}{2}$	$\frac{1}{2}$
AAAA	0	0	1

Some initial population G0 produces gametes with haplotypic composition:

Haplotype:

$$aa \quad Aa \quad AA$$
$$f \quad g0,1 \quad g1,1 \quad g2,1$$

The frequency of a is,

$$q = g0,1 + \frac{1}{2}g1,1$$

And that of A is,

$$p = \frac{1}{2}g1,1 + g2,1$$

Panmictic reproduction of G_0 yields population G_1 with the following genotypic composition:

Genotype:

$$aaaa \quad Aaaa \quad AAaa \quad AAAa \quad AAAA$$
$$f \quad g0,1^2 \quad 2g0,1g1,1 \quad g1,1^2 + 2g0,1g2,1 \quad 2g1,1g2,1 \quad g2,1$$

The haplotypic composition of the gametes produced by G_1 is:

Haplotype:

$$aa \quad Aa \quad AA$$
$$f \quad g0,2 \quad g1,2 \quad g2,2$$

According to table the following applies:

$$g1,2 = \frac{1}{2}(2g0,1g1,1) + \frac{2}{3}(g1,1^2 + 2g0,1g2,1) + \frac{1}{2}(2g1,1g2,1)$$

$$= \frac{2}{3}(\frac{3}{2}g0,1g1,1 + \frac{3}{2}g1,1g2,1 + g1,1^2 + 2g0,1g2,1)$$

$$= \frac{2}{3}[2(g0,1 + \frac{1}{2}g1,1)(\frac{1}{2}g1,1 + g2,1) + \frac{1}{2}g1,1(g0,1 + g1,1 + g$$

$$= \frac{2}{3}(2pq + \frac{1}{2}g1,1)$$

Generally

$$g1,t+1=\frac{2}{3}(2pq+\frac{1}{2}g1,t)$$

The frequencies of the genotypes have attained their equilibrium (*e*) values as soon as the frequencies of the haplotypes are constant. *The latter implies*:

$$g1,e=\frac{2}{3}(2pq+\frac{1}{2}g1,e),$$

i.e.

$$g1,e=2pq$$

The haplotype frequencies are then,

$$g0,e=q-\frac{1}{2}g1,e=q-pq=q^2$$

$$g1,e=2pq$$

$$g2,e=p-\frac{1}{2}g1,e=p-pq=p^2$$

The genotypic composition in equilibrium is consequent:
Genotype:

	aaaa	Aaaa	AAaa	AAAa	AAAA
f	q^4	$4pq^3$	$6p^2q^2$	$4p^3q$	p^4

This composition is also given by the probability distribution for J, the number of A alleles in the autotetraploid genotype:

$$P(\underline{j}=J)=\binom{4}{J}p^Jq^{4-J}$$

The deviation from the equilibrium is measured by the quantity *dt*, which measures the excess or deficit of the frequency of gametes with the *Aa* haplotype with regard to their equilibrium frequency.

Thus dt is defined as follows:

$$d_t:=g1,t-g1,e$$

The rate of decrease of *dt* indicates how fast the equilibrium is approached.

Equations Yield:

$$d_{t+1}=g1,t+1-g1,e=\frac{2}{3}(2pq+\frac{1}{2}g1,t)-2pq=\frac{1}{3}(g1,t-g1,e)=\frac{1}{3}dt$$

One round of panmictic reproduction produces a population in which the deviation amounts only to 1/3 of the deviation in the preceding population. The equilibrium is approached in an asymptotic way. Example gives an illustration.

Example: The approach of the equilibrium is considered for an initial population G0 with genotypic composition (0.04; 0; 0.72; 0; 0.24).

The haplotype frequencies are:

$$g0,1 = 0.04 + 0.12 = 0.16$$
$$g1,1 = 0.48$$
$$g2,1 = 0.12 + 0.24 = 0.36$$

Thus q = 0.4 and p = 0.6.

This implies that:

$$g0,1 = q^2 = g0,e$$
$$g1,1 = 2pq = g1,e$$
$$g2,1 = p^2 = g2,e$$

For a more advanced treatment of the population genetic theory of crossfertilizing crops with an autotetraploid behaviour of the chromosomes the reader is referred to Seyffert. Finally, it is emphasized once again that in this section it was assumed that the population contains only two different alleles for the segregating locus.

In fact more alleles may occur in such a way that plants with three or four different alleles per locus are present, *viz*. plants with genotype *AiAiAjAk* or *AiAjAkAl*, respectively.

Quiros reported such genotypes for isozyme loci in alfalfa. Some claims have been made that plants with a heterozygous genotype containing three or four different alleles for the considered locus, are more vigourous than plants with a heterozygous genotype containing one or two alleles.

Genetic Effects of Inbreeding

Inbreeding occurs if mating plants are, on the average, *more* related than random pairs of plants. A more than average relatedness of the mating plants is thus a prerequisite. Relatedness implies, of course, that the plants involved share one or more ancestors. The strength of the inbreeding depends on the degree of relatedness of the mating plants. It has already been noted that mating of related plants may occur in random mating, but in that case it occurs as a matter of chance.

Note: Several yardsticks for measuring the degree of relatedness exist, a common one being the probability that an allele of a certain locus in some plant is identical by descent to an arbitrary allele at that same locus in its mate. In regular systems of inbreeding the degree of relatedness of the mating plants is uniform across all pairs of mating plants. In this book no attention is given to the determination of the degree of relatedness.

Regular systems of inbreeding are far more common in plant breeding than irregular systems. No attention will, therefore, be given to irregular systems of inbreeding. The counterpart of inbreeding is outbreeding. With outbreeding mating plants are on the average *less* related than random pairs of plants. Selfincompatibility is a natural cause for outbreeding as related plants tend to have a similar genotype at the incompatibility locus/loci. After intercrossing, such plants will produce no (or few) offspring.

Artificial forms of outbreeding are:

• Bulk crossing of two unrelated populations
• Selection of parents to be crossed in such a way that inbreeding is avoided as much as possible Outbreeding occurs also in the case of immigration.

The population genetic effect of inbreeding is a decrease in the frequency of heterozygous plants. This involves all loci, for all traits. (Random mating, on the other hand, is a mode of reproduction that may occur for certain traits and may simultaneously be absent for other traits). When starting with an F2 population and considering segregating loci, the frequency of heterozygous plants is the same for all loci. This applies to the successive generations of the superpopulation. Each subpopulation consists of few plants: in the case of selfing only a single plant, in the case of full sib mating only pairs of plants. Within these separate subpopulations reproduction is by means of random mating.

The random variation of the gene frequencies occurring in small populations causes the subpopulations to vary with regard to the frequencies of heterozygous plants: not only for different loci, but also for the same locus. Individual plants of the F2 (or F3, etc.) populations vary therefore in the number of heterozygous loci. In diploid crops procedures for the production of doubled haploid lines (DH-lines) allow the production of pure lines from heterozygous parents in a single generation.

Doubling of the number of chromosomes of haploid plants, generated by parthenogenesis or by anther culture, yields immediately complete homozygosity. For dioecious crops as well as for self-fertilizing crops with a long juvenile phase, *e.g. Coffea arabica* L., this approach is an attractive alternative to continued inbreeding. Tissue culture techniques for the regeneration of plants from anthers or microspores have been developed, for example in wheat, barley, rice and oilseed rape. Also elimination of paternal chromosomes, occurring when making *Hordeum vulgare* L. × *H. bulbosum* L. or *Triticum aestivum* L. × *Zea mays* L. crosses, permits production of DH-lines. (The paternal chromosomes are lost in a few cell divisions of the hybrid zygote/ embryo.)

Note: DH-lines are mostly obtained directly from the gametes produced by the F1-plants. This has a few drawbacks

- Recombination is restricted to the F1 meiosis
- The proportion of DH-lines that are rejected because of poor performance is high. This is undesirable because of the cost of producing DH-lines.

To avoid these drawbacks one may use gametes from plants obtained by backcrossing the F1 or one may use F2- or even F3-plants. (The latter allows selection among F2-plants, followed by selection among F3-lines in the seedling stage). *In vitro* selection among the haploid embryos appeared to be feasible: the size and degree of embryo differentiation predicted which embryos would produce vigourous seedlings. Additionally the growth rate of the embryos was positively correlated with yield performance in the field $r = 0.3$, but this has found little practical application).

Continued self-fertilization is the natural mode of reproduction of selffertilizing crops. There are many economically important self-fertilizing crops.

A number of these are:

- Barley *Hordeum vulgare* L.
- Oats *Avena sativa* L.
- Wheat *Triticum aestivum* L.
- Rice *Oryza sativa* L.

- Sorghum *Sorghum bicolour* (L.) Moench.
- Finger millet *Eleusine coracana* (L.) Gaertn.
- Pea *Pisum sativum* L.
- Cowpea *Vigna unguiculata* (L.) Walp.
- Dry bean *Phaseolus vulgaris* L.
- Soybean *Glycine max* (L.) Merr.
- Peanut *Arachis hypogaea* L.
- Cotton *Gossypium* spp.
- Arabica coffee *Coffea arabica* L.
- Lettuce *Lactuca sativa* L.
- Tomato *Lycopersicon esculentum* Mill.
- Okra *Abelmoschus esculentus* (L.) Moench.
- Sweet pepper *Capsicum annuum* L.

Self-fertilization is not always 100% in most of these autogamous crops, *e.g.* cotton, okra, sorghum. (The amount of outcrossing in sorghum is about 6%.) consider the genotypic composition of populations reproducing by a mixture of self-fertilization and cross-fertilization. Breeders regularly apply inbreeding in cross-fertilizing crops.

They may have various reasons for doing this:
- The development of pure lines (mostly by continued selfing) for use as parents in the breeding of hybrid varieties, *e.g.* in maize or cucumber.
- To promote the efficiency of elimination of an undesired recessive gene
- Maintenance of a genic male sterile 'line'

Note:FS-mating occurs also when a maintaining a genic male sterile barley 'line': male sterile plants are harvested after having been pollinated by their male fertile full sibs. (This is also applied in the case of recurrent selection in self-fertilizing cereals).

Thus the harvesting of a female plant (say genotype mm) implies harvest of seed due to the cross $mm \times Mm$ (where Mm represents the genotype assumed for hermaphroditic plants). The genotypic composition of the obtained FS-family is (1/2, 1/2, 0). Repeated application of this procedure implies repeated FSmating.

The most powerful form of inbreeding of cross-fertilizing crops, *e.g.* dioecious crops, occurs with repeated crossing of the type:

- Full sib × full sib, *i.e.* full sib mating, or
- Parent ×× offspring.

Full Sib Mating

The offspring due to a cross of two genotypes constitutes a family. The plants belonging to the family share both their maternal and their paternal parent. With regard to each other these plants are full sibs. Together they form a full sib family (FS-family). Crossing of plants belonging to the same FS-family is called full sib mating (FS-mating). FS-mating may be used when inbreeding of dioecious crops, such as spinach or asparagus, is the aim. It occurs spontaneously in the case of open pollination within FS-families grown in isolation. This is applied in hermaphroditic, monoecious or dioecious crops in the case of separated FS-family selection. Note describes how FS-mating is applied when maintaining a genic male sterile 'line'.

Parent ×× offspring mating

In this book the notation A ×× B indicates the cross A×B and/or the reciprocal cross B × A. Parent ×× offspring crosses, *i.e.* so-called PO-mating, can only be applied to perennial crops such as oil palm (producing gametes from the age of 4–5 years for many years) or asparagus (with a juvenile phase lasting two years).

The parent is still alive when its offspring reach the reproductive phase. Repeated backcrossing implies continued application of crosses of the type 'recurrent parent ×× offspring'. In the absence of selection the genotype of the offspring becomes identical to the genotype of the recurrent parent (if the recurrent parent has a homozygous genotype) or to the genotypic composition of the possible lines obtained by selfing of the recurrent parent (if the recurrent parent is heterozygous). In this chapter only loci segregating for not more than two alleles per locus will be considered. For an extensive treatment of the population genetics theory of inbreeding the reader is referred to Allard, Jain and Workman.

Diploid Chromosome Behaviour and Inbreeding

One Locus with Two Alleles

With continued inbreeding of any (infinitely) large population the genotype frequencies will change from one generation to the other until the frequency of plants with a heterozygous genotype has become zero. Starting from the initial population G0 with genotypic composition $(f0,0, f1,0, f2,0)$, eventually a population with genotypic composition $(q, 0, p)$ will be obtained.

- Starting with some arbitrary genotypic composition

Genotype:

Generation	aa	Aa	AA
S0	f0	f1	f2
S1	$f0 + \dfrac{1}{4}f1$	$\dfrac{1}{2}f1$	$f2 + \dfrac{1}{4}f1$
S2	$f0 + \left(\dfrac{1}{4} + \dfrac{1}{8}\right)f1$	$\dfrac{1}{4}f1$	$f2 + \left(\dfrac{1}{4} + \dfrac{1}{8}\right)f1$
S3	$f0 + \left(\dfrac{1}{4} + \dfrac{1}{8} + \dfrac{1}{16}\right)f1$	$\dfrac{1}{8}f1$	$f2 + \left(\dfrac{1}{4} + \dfrac{1}{8} + \dfrac{1}{16}\right)f1$
.			
.			
S∞	q	0	P

- Starting with F1, *i.e.* a population with genotypic composition (0, 1, 0)

Generation (t)	Population	Inbreeding Coefficient	Panmictic ($\wp$index(P)	Genotype aa	Aa	AA
0	$S_0(=F_1)$	−1	2	0	1	0
1	$S_1(=F_2)$	0	1	$\dfrac{1}{4}$	$\dfrac{1}{2}$	$\dfrac{1}{4}$
2	$S_2(=F_3)$	$\dfrac{1}{2}$	$\dfrac{1}{2}$	$\dfrac{3}{8}$	$\dfrac{2}{8}$	$\dfrac{3}{8}$
3	$S_3(=F_4)$	$\dfrac{3}{4}$	$\dfrac{1}{4}$	$\dfrac{7}{16}$	$\dfrac{2}{16}$	$\dfrac{7}{16}$
4	$S_4(=F_5)$	$\dfrac{7}{8}$	$\dfrac{1}{8}$	$\dfrac{15}{32}$	$\dfrac{2}{32}$	$\dfrac{15}{32}$
5	$S_5(=F_6)$	$\dfrac{15}{16}$	$\dfrac{1}{16}$	$\dfrac{31}{64}$	$\dfrac{2}{64}$	$\dfrac{31}{64}$
6	$S_6(=F_7)$	$\dfrac{31}{32}$	$\dfrac{1}{32}$	$\dfrac{63}{128}$	$\dfrac{2}{128}$	$\dfrac{63}{128}$
7	$S_7(=F_8)$	$\dfrac{63}{64}$	$\dfrac{1}{64}$	$\dfrac{127}{256}$	$\dfrac{2}{256}$	$\dfrac{127}{256}$
∞	$S_\infty(=F_\infty)$	1	0	$\dfrac{1}{2}$	0	$\dfrac{1}{2}$

Illustrates this for inbreeding by means of continued selfing. It appears that the genotype frequencies approach, in an asymptotic manner, the gene and haplotype frequencies. Often the frequency of heterozygous plants in generation t, *i.e.* $f1,t$, is written in the form,

$$2pq(1 - f_t)$$

In this expression the factor $1-Ft$ describes the deviation from the Hardy–Weinberg frequency. The factor is called the panmictic index, sometimes designated by the symbol P. This implies that $P = 1-Ft$. The parameter Ft, say 'script F', is the inbreeding coefficient (or fixation index) pertaining to generation t. When starting with an F1 population, F2 is the first generation due to self-fertilization. For this reason the F2 population is chosen to be generation 1. (Its genotypic composition is equal to the genotypic composition of the population obtained by panmictic reproduction of the F1) Successive generations may be indicated by G1,G2..., but in the case of continued selfing the designations S1, S2, S3... are used as well. A general description of the genotypic composition of any population (inbred or not) is now given by,

Genotype

	aa	Aa	AA
f	$q^2 + pqf_t$	$2pq(1-f_t)$	$P^2 + pqf_t$

In several other books, e.g. Falconer and MacKay, the inbreeding coefficient is defined as the probability that the two alleles at any loci of a plant are identical by descent. This would mean that the inbreeding coefficient of an F2 population obtained from cross $AA \times aa$ is equal to 1 2, because 50% of the plants contain, for locus A-a, alleles that are identical by descent (this concerns plants with genotype aa or AA). In this book the parameter F is used to quantify the deviations from the Hardy–Weinberg frequencies. In an F2 population such deviations are absent and accordingly its inbreeding coefficient is 0. In Note, it is shown that our definition of the inbreeding coefficient F can be interpreted as the coefficient of correlation of numerical values, e.g. gene-effects, assigned to the haplotypes of the uniting gametes.

This is based on the following consideration. With random mating the gene effects of the haplotypes of fusing female and male gametes are independent; in the absence of random mating they are interdependent. With inbreeding they tend to be similar; with outbreeding they tend to be different. Breeding of self-fertilizing crops starts mostly with crossing of homozygous lines. For all loci for which the parental lines have a different homozygous genotype the genotype of the F1 is heterozygous. For these loci $p = q = 1$ 2 and then the expressions in (3.1) simplify to

Genotype

	aa	Aa	AA
f	$\frac{1}{4}(1+f_t)$	$\frac{1}{2}(1-f_t)$	$\frac{1}{4}(1+f_t)$

As $f1,0 = 1\ 2\ (1-F0) = 1$, it follows that $F0 = -1$, *i.e.* a negative value for the inbreeding coefficient. The panmictic index of the F1 amounts for heterozygous loci to $P0 = 2$. In the remainder of this section the decrease in the frequency of heterozygous plants is considered for the three most important regular inbreeding systems, *viz.* self-fertilization, full sib mating and parent × offspring mating. To measure this decrease the parameter λ is defined:

$$\lambda = \frac{2pq(1-f_t)}{2pq(1-f_{t-1})} = \frac{1-f_t}{1-f_{t-1}}$$

This parameter indicates the frequency of heterozygous plants as a proportion of this frequency in the preceding generation. At a smaller value for λ the decrease of $f1$ is stronger. In the case of selfing the values for λ do not depend on t; they are approximately constant when applying full sib mating or parent × offspring. Then $\lambda1 = \lambda2 = \cdots = \lambda t$. This implies,

$$f1,t = \lambda f_{1,t-1} = \lambda^2 f1,t - 2 = \lambda^t f1,0$$

Self-Fertilization

In the F2 generation, the first generation generated by selfing, the genotype frequencies coincide with the Hardy-Weinberg frequencies. Thus $f1,1 = 2pq$, implying that $F1$, the inbreeding coefficient of F2, is zero. In population F−, approximately obtained after a very large number of generations reproducing by means of selfing, there is complete homozygosity, *i.e.* $f1,- = 0$, implying that $F-$, the inbreeding coefficient of F−, is 1. The decrease of $f1$, due to continued selfing, is indicated in Table. The table shows that $f1$ is halved by each round of reproduction by means of selfing.

Thus,

$$1 - f_t = \frac{1}{2}(1 - f_{t-1})$$

implying

$$f_t = \frac{1}{2}(1 + f_{t-1})$$

With Regard to Continued selfing the expression:

$$1 - f_t = \frac{1}{2}(1 - f_{t-1})$$

or

$$P_t = \frac{1}{2}P_{t-1}$$

implies

$$P_t = \left(\frac{1}{2}\right)^t P_0 = \left(\frac{1}{2}\right)^{t-1}$$

$$f_t = 1 - \left(\frac{1}{2}\right)^{t-1}$$

At all other systems of inbreeding the reduction of $f1$ is smaller. The minimum value for l is thus attained with selfing. It amounts to lS = 1/2.

Full Sib Mating and Parent Offspring Mating

Li showed that for both full sib mating and parent × offspring mating, the relation:

$$f1,t+2 = \frac{1}{2}f1,t+1 + \frac{1}{4}f1,t$$

Applies. Consider an initial population with genotypic composition (0,1,0), thus $f1,0 = 1$. In this population plants are crossed in pairwise combinations. In the next generation the genotypic composition of the population obtained, which consists of full sib families, is expected to be (1/2, 1/4, 1/2), with $f1,1 = 1\ 2$.

Continued full sib mating, within the continuously generated FS-families, gives, according to Equation,

$$f1,2 = \frac{1}{2}\left(\frac{1}{2}\right) + \frac{1}{4}(1) = \frac{1}{2}, i.e. \lambda_2 = 1$$

$$f1,3 = \frac{1}{2}\left(\frac{1}{2}\right) + \frac{1}{4}\left(\frac{1}{2}\right) = \frac{3}{8}, i.e. \lambda_3 = \frac{3}{4} = 0.75$$

$$f1,4 = \frac{1}{2}\left(\frac{3}{8}\right) + \frac{1}{4}\left(\frac{1}{2}\right) = \frac{5}{16}, i.e. \lambda_4 = \frac{5}{6} = 0.8333, etc$$

The first round of inbreeding (full sib mating or parent × offspring mating) does not give a decrease of the frequency of heterozygous plants ($\lambda 2 = 1$). Indeed, with full sib mating first FS-families have to be generated. It appears that λ approaches asymptotically the value $\lambda FS = \lambda PO = 0.809$. As $(0.809)3 = 0.53 \approx 1/2$, three generations of reproduction by means of FSmating or parent × offspring mating give the same reduction in $f1$ as a single round of reproduction by selfing.

A pair of linked loci

It was shown that linkage may be expected to play a relatively unimportant role in the inheritance of quantitative traits. It was said that, throughout this book, absence of linkage would be assumed. It is, nevertheless, useful to be familiar with some implications of linkage. An important reason for this is the study of the linkage of loci affecting a quantitative trait with molecular markers. Consider haplotypes ab, aB, Ab or AB for the two loci A-a and B-b with recombination value rc. Continued selfing, starting with an F1 with the heterozygous genotype $AaBb$, yields in the absence of selection

$$g11,t = g00,t$$
$$g01,t = g10,t$$

'symmetric' haplotype frequencies:

$$g11,t + g10,t = PA = \frac{1}{2}$$

$$g10,t = \frac{1}{2} - g11,t$$

This implies that, when one knows $g11,t$, one also knows $g10,t$, $g01,t$ and $g00,t$. It suffices thus to consider only the frequency of gametes with the AB haplotype. This is particularly of interest when considering F–. This population is described by

Genotype

aabb	AAbb	aaBB	AABB
f $f00,\infty$	$f20,\infty$	$f02,\infty$	$f22,\infty$

Only plants with the $AABB$ genotype are capable of producing gametes with the AB haplotype. Thus,

$$g11,- = f22,-.$$

The haplotypic composition of the gametes produced by this population is

Haplotype

ab	Ab	aB	AB
g $g00,\infty(=g11,\infty)$	$g10,\infty(=\frac{1}{2}-g11,\infty)$	$g01,\infty(\frac{1}{2}-g11,\infty)$	$g11,\infty$

There are thus good reasons to consider the frequency of gametes with the *AB* haplotype. In Note the following relation between the frequencies of *AB*-haplotypes in two successive generations is derived:

Note: The frequency of *AB* haplotypes, *i.e.* g11, is considered for the case of continued autogamous reproduction. (To promote readability the recombination value is – in this section – mostly just indicated by the symbol *r*).

The genotypes capable of producing *AB* haplotypes, their frequencies in generation *t* and the haplotypic composition of the gametes they produce are:

Haplotype:

Genotype	f	ab	aB	Ab	AB
AABB	f22,t	0	0	0	1
AABb	f21,t	0	0	$\frac{1}{2}$	$\frac{1}{2}$
AaBB	f12,t	0	$\frac{1}{2}$	0	$\frac{1}{2}$
AB/ab	f11c,t	$\frac{1}{2}(1-r)$	$\frac{1}{2}r$	$\frac{1}{2}r$	$\frac{1}{2}(1-r)$
Ab/aB	f11R,t	$\frac{1}{2}r$	$\frac{1}{2}(1-r)$	$\frac{1}{2}(1-r)$	$\frac{1}{2}r$

Then

$$g_{11,t+1} = f_{22,t} + \frac{1}{2}f_{21,t} + \frac{1}{2}f_{12,t} + \frac{1}{2}(1-r)f_{11C,t} + \frac{1}{2}rf_{11R,t}$$

$$= f_{22,t} + \frac{1}{2}f_{21,t} + \frac{1}{2}f_{12,t} - \frac{1}{2}r(f_{11C,t} - f_{11R,t})$$

$$= f_{22,t} + \frac{1}{2}f_{21,t} + \frac{1}{2}f_{12,t} + \frac{1}{2}f_{11C,t} - rd_t$$

Where, according to equation d_t is defined as,

$$d_t = \frac{1}{2}(f11c,t - f11R,t)$$

and

$$f22,t = f22,t-1 + \frac{1}{4}f21,t-1 + \frac{1}{4}f12,t-1 + \frac{1}{4}(1-r)^2 f11c,t-1 + \frac{1}{4}r^2 f11R,t-1$$

$$f21,t = \frac{1}{2}f21,t-1 + \frac{1}{2}r(1-r)f11c,t-1 + \frac{1}{2}r(1-r)f11R,t-1$$

$$f12,t = \frac{1}{2}f12,t-1 + \frac{1}{2}r(1-r)f11C,t-1 + \frac{1}{2}r(1-r)f11R,t-1$$

$$f11C,t = \frac{1}{2}(1-r)^2 f11C,t-1 + \frac{1}{2}r^2 f11R,t-1$$

$$f11R,t = \frac{1}{2}r^2 f11C,t-1 + \frac{1}{2}(1-r)^2 f11R,t-1$$

Thus,

$$g11,t+1 = f22,t-1+\left(\frac{1}{4}+\frac{1}{4}\right)f21,t-1+\left(\frac{1}{4}+\frac{1}{4}\right)f12,t-1+[\frac{1}{4}(1-r)^2$$

$$+\frac{1}{4}r(1-r)+\frac{1}{4}r(1-r)+\frac{1}{4}(1-r)^2]f11C,t-1$$

$$+[\frac{1}{4}r^2+\frac{1}{4}r(1-r)+\frac{1}{4}r(1-r)+\frac{1}{4}r^2]f11R,t-1^{-rdt}$$

$$= f22,t-1+\frac{1}{2}f21,t-1+\frac{1}{2}f12,t-1+\frac{1}{2}(1-r)f11C,t-1$$

$$+\frac{1}{2}rf11R,t-1^{-rdt}$$

$$= g11,t-rd_t$$

(This equation is identical to Equation derived for the case of continued panmictic reproduction.)

$$g11,t+1 = g11,t - r_c d_t$$

Equation applies at continued self-fertilization. It is identical to Equation applying at continued panmictic reproduction. One should realise, however, that with panmictic reproduction the relation between $dt+1$ and dt was derived to be $d_{t+1}=(1-r_c)d_t$. For autogamous reproduction, however, the relation between dt and $dt-1$ can be shown to be

$$d_{t+1}=\left(\frac{1-2r_c}{2}\right)d_t$$

Note: In the case of (continued) selfing, plants with a doubly heterozygous genotype, in the coupling phase or in the repulsion phase, can only be produced by doubly heterozygous parents, one can easily derive from Table that:

$$f_{11C,t+1}=2\left(\frac{1-r}{2}\right)^2 f_{11c,t}+2\left(\frac{r}{2}\right)^2 f_{11R,t}$$

$$f_{11R,t+1}=2\left(\frac{1-r}{2}\right)^2 f_{11R,t}+2\left(\frac{r}{2}\right)^2 f_{11C,t}$$

Thus,

$$f_{11,t+1}=\left[2\left(\frac{1-r}{2}\right)^2+2\left(\frac{r}{2}\right)^2\right](f_{11C,t}+f_{11R,t})$$

$$=\left(r^2-r+\frac{1}{2}\right)f_{11,t}=\left[\left(r-\frac{1}{2}\right)^2+\frac{1}{4}\right]f_{11,t}$$

Equations:

$$d_{t+1} = \frac{1}{2}\left(f_{11,Ct+1} - f_{11,Rt+1}\right)$$

yield thus,

$$d_{t+1} = \frac{1}{4}\left[(1-r)^2 - r^2\right]\left(f_{11,Ct+1} - f_{11,Rt+1}\right)$$

This gives Equation,

$$d_{t+1} = \left(\frac{1-2r_c}{2}\right)d_t$$

$$d_t = \left(\frac{1-2r_c}{2}\right)^{t-1}d_t$$

Equations yield for the case of continued selfing:

$$g_{11,t+1} = g_{11,t} - r_c\left(\frac{1-2r_c}{2}\right)d_{t-1}$$

The parameter dt is still, as defined in Equation, equal to 1/2 $(f11C,t-f11R,t)$. Equation shows that, unless $dt = 0$ or $rc = 1/2$, the haplotype frequencies will change from one generation to the next. The genotypic composition of F–, for F1 in coupling phase as well as in repulsion phase, depends directly on Equation, *viz.*

$$g_{11,\infty} = f_{22,\infty} = g_{11,1} - \left(\frac{2r}{1+2r}\right)d_1$$

Note: Combining the Equation yields in the case of continued selfing

$$g_{11,t+1} - g_{11,t} = -rd_1\left(\frac{1-2r}{2}\right)^{t-1}$$

Repeated application of this equation result via:

$$g_{11,2} - g_{11,1} = -rd_1\left(\frac{1-2r}{2}\right)^0$$

$$g_{11,3} - g_{11,2} = -rd_1\left(\frac{1-2r}{2}\right)^1$$

$$g_{11,t+1} - g_{11,t} = -rd_1\left(\frac{1-2r}{2}\right)^{t-1}$$

$$g_{11,t+1} - g_{11,1} = -rd_1\sum_{j=0}^{t-1}\left(\frac{1-2r}{2}\right)^j$$

The sum of the terms of this geometric series is:

$$\frac{1-\left(\frac{1-2r}{2}\right)^{t-1}}{1-\left(\frac{1-2r}{2}\right)} = \frac{2}{1+2r}\left[1-\left(\frac{1-2r}{2}\right)^{t-1}\right]$$

Thus,

$$g_{11,t+1} = g_{11,1-r}\left(\frac{2}{1+2r}\right) \cdot d_1 \cdot \left[1-\left(\frac{1-2r}{2}\right)^{t-1}\right]$$

Implying,

$$g_{11,\infty} = f_{22,\infty} = g_{11,1} - \left(\frac{2r}{1+2r}\right)d_1$$

The quantity to be substituted in Equation for $d1$ amounts, according to Example, to $1/4$ $(1-2r)$ for F1 in the coupling phase and to $-1/4$ $(1-2r)$ for F1 in the repulsion phase.

Equation yields thus for F1 in the coupling phase:

$$g_{11,\infty} = f_{22,\infty} = \left(\frac{1-r}{2}\right) - \left(\frac{2r}{1+2r}\right)\left(\frac{1-2r}{4}\right) = \frac{1}{2(1+2r)}$$

For F_1 in the repulsion phase we get,

$$g_{11,\infty} = f_{22,\infty} = \left(\frac{r}{2}\right) + \left(\frac{2r}{1+2r}\right)\left(\frac{1-2r}{4}\right) = \frac{2r}{2(1+2r)}$$

The genotypic composition of F– with regard to the complex genotype s for the two linked loci A-a and B-b.

- F_1 in coupling phase

	$\frac{bb}{2rc}$ $\frac{}{2(1+2r_c)}$	Bb	BB	
aa	0	0	$\frac{1}{2(1+2r_c)}$	$\frac{1}{2}$
Aa	$\frac{1}{2(1+2r_c)}$	0	0	0
AA	$\frac{1}{2}$	0	$\frac{2rc}{2(1+2r_c)}$	$\frac{1}{2}$
		0	$\frac{1}{2}$	1

- F_1 in repulsion phase

	bb $\dfrac{2rc}{2(1+2r_c)}$	Bb	BB	
aa	0	0	$\dfrac{1}{2(1+2r_c)}$	$\dfrac{1}{2}$
Aa	$\dfrac{1}{2(1+2r_c)}$	0	0	0
AA	$\dfrac{1}{2}$	0	$\dfrac{2rc}{2(1+2r_c)}$	$\dfrac{1}{2}$
	0	$\dfrac{1}{2}$	1	

Table presents the genotypic composition of F−. It may be compared with Table presenting the genotypic composition obtained after continued panmixis.

In the case of linkage ($0 < rc < 1/2$) the frequencies of the haplotypes change in the course of the generations. For gametes with the AB haplotype the difference between $g11,1$ and $g11,-$ amounts to,

$$g_{11,\infty} - g_{11,1} = \left(\frac{2r}{1+2r}\right) d_1$$

This amounts, according to Example, for F1 in the coupling phase to,

$$\left(\frac{2r}{1+2r}\right)\left(\frac{1-2r}{4}\right) = \frac{r(1-2r)}{2(1+2r)}$$

And for F_1 in the repulsion phase to,

$$\left(\frac{2r}{1+2r}\right)\left(\frac{2r-1}{4}\right) = \frac{r(2r-1)}{2(1+2r)}$$

These differences are for $0 < rc < 1$ 2 generally quite small. For $rc = 1/4$, for instance, it amounts for F1 in the repulsion phase to $g11,1$-$g11,-$ = 1/8-1 6 = −0.0417. We consider now the frequency of plants with a genotype obtained by crossing two parents.

It may, for example, be desired to obtain genotype $AABB$ from an initial cross of genotypes $AAbb$ and $aaBB$. The frequency of $AABB$ plants amounts in population F2 to $f22,1$ = 1/4 $rc/2$. Equation

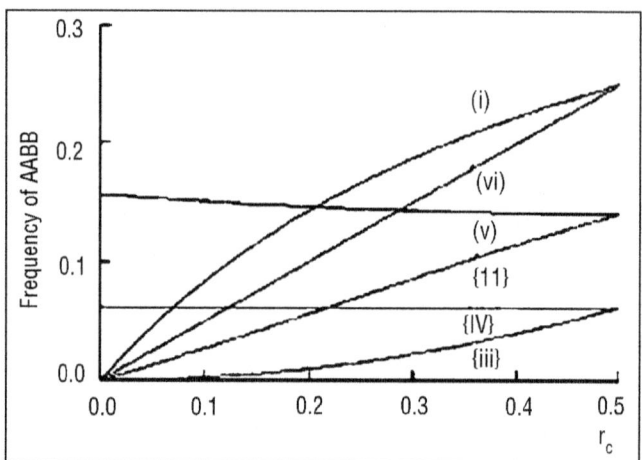

Figure 1: The Frequency of Plants with Genotype AABB as a Function of the Recombination Value rc.

Considered are population obtained by crossing of genotypes AAbb and aaBB followed by:

- Continued self-fertilization until F-,
- Selfing until F3,
- Selfing until F_2,
- Continued panmixis until linkage equilibrium,
- Continued panmixis followed by one round of reproduction by means of selfing, or
- Doubling of the number of chromosomes in the gametes produced by F_1 yields for $t = 2$ the frequency of plants with genotype *AABB* in F3.

When substituting the F2 genotype frequencies presented in Table one gets for an F1 in the repulsion phase:

$$f22,2 = \frac{1}{4}r^2 + \frac{1}{8}r(1-r) + \frac{1}{8}r(1-r) + \frac{1}{8}r^2(1-r)^2 + \frac{1}{8}(1-r)^2r^2$$

$$= \frac{1}{4}r + \frac{1}{4}r^2 - \frac{1}{2}r^3 + \frac{1}{4}r^4$$

This amounts, for unlinked loci, to $f22,2 = 9\ 64 = \_3/8\ \_2 = f00,2$. According to Equation the frequency of *AABB* plants in F– is $2r$ $2(1+2r)$. Because $2r\ 2(1+2r) \leq 1/2(1+2r)$, plants with one of the parental genotypes will outnumber plants with this recombinant genotype to a greater extent as linkage is stronger, *i.e.* as rc is smaller.

In populations of cross-fertilizing crops, doubly heterozygous genotypes tend to be permanently present; in populations of self-fertilizing crops they disappear.

One should, however, be careful when speaking about 'the recombining effect of cross-fertilization'. This is illustrated for loci A-a and B-b. Continued panmictic reproduction gives eventually, at linkage equilibrium, $f22 = p2r2$. This amounts for $p = r = 1/2$ to 1/16, whatever the recombination value. For tightly linked loci, with $rc < 1\ 14$, genotype $AABB$ will indeed occur with a higher frequency in populations in linkage equilibrium than in populations obtained by continued selfing.

For less tightly linked loci, i.e. $rc > 1/14$, the frequency of $AABB$ will, however, be higher in F–. Thus one should not decide rashly to increase the frequency of plants with a recombinant genotype by the application of random mating in F2, F3... populations of a self-fertilizing crop. With regard to unlinked loci continued random mating will only result in the genotypic composition of F2, because for unlinked loci the $F2$ population obtained by selfing will have the linkage equilibrium composition. Selection in a cross-fertilizing crop is more efficient when increasing the frequency of homozygous recombinant genotypes by selfing. According to Note a single round of reproduction by means of self-fertilization in a population in linkage equilibrium gives,

$$f_{22} = \frac{5 - 2r + 2r^2}{32}$$

Note: Consider a population in linkage equilibrium. It is obtained by panmictic reproduction starting with a single-cross hybrid variety. With regard to loci A-a and B-B a single round of reproduction by means of selfing results, according to Equation, in the following frequency of plants with genotype $AABB$:

$$f_{22} = \frac{1}{16} + \frac{1}{4}.\frac{1}{8} + \frac{1}{4}.\frac{1}{8} + \frac{1}{4}r^2.\frac{1}{8} + \frac{1}{4}(1-r)^2.\frac{1}{8} = \frac{5 - 2r + 2r^2}{32}$$

For $r = 1/2$ this amounts to 9 64, i.e. _3/8 _2. It is the same value as obtained, from Equation, for an F3. The single reproduction by means of selfing gives thus the genotypic composition of an F3. This illustrates that the genotypic composition of the population in linkage equilibrium is equal to the genotypic composition for pairs of unlinked loci in an F2. In a diploid crop, doubling the number of chromosomes of haploid plants is the fastest way to attain complete homozygosity.

The frequency of plants with the desired recombinant genotype then amounts to $1/2$ rc, $i.e.$ $2/rc$ times as high as in F2. The frequency of doubly heterozygous plants is greatly reduced with reproduction by means of selfing. Depending on the recombination value, a single round of selfing reduces this frequency to only $1/4$ to $1/2$ of the frequency of plants with the $AaBb$ genotype in the preceding generation. Note shows that the remaining portion of doubly heterozygous plants amounts to $f11,t+1$ $f11,t = (r-1/2)2 + 1$ 4, which amounts to $1/4$ for $rc = 1/2$ and to $1/2$ for $rc = 0$. This reduction of the frequency of heterozygous plants is even stronger for more complex genotypes: a single round of selfing reduces the frequency of the complex genotype consisting of a heterozygous single-locus genotype for each of k unlinked loci to the portion $(1/2)k$ of its preceding value.

Autotetraploid Chromosome Behaviour and Self-Fertilization

Spontaneous self-fertilization as the natural mode of reproduction occurs rather rarely among crops with an autotetraploid chromosome behaviour. The somatic chromosome number of quinoa (*Chenopodium quinoa*) is $2n = 36$. The basic chromosome number for the genus *Chenopodium* is $x = 9$. This suggests that quinoa is a tetraploid. Ward found for the same locus both diploid and tetraploid behaviour. Simmonds reported that selfing predominates, without evident inbreeding depression.

Quite a few autotetraploid crops, $e.g.$ durum wheat (*Triticum durum*; $2n = 4x = 28$) or coffee (*Coffea arabica*; $2n = 4x = 44$), have a diploid chromosome behaviour. For other crops, $e.g.$ European potato (*Solanum tuberosum*; $2n = 4x = 48$) or wild barley (*Hordeum bulbosum*; $2n = 4x = 28$), there may be a more or less perfect autotetraploid chromosome behaviour, implying that exclusively quadrivalents are being formed at meiosis. Artificial selffertilization may be applied in a man-made autotetraploid crop such as rye (*Secale cereale*; $2n = 4x = 28$), which is self-incompatible in its natural diploid condition.

In this section attention is only given to the simple situation of a single segregating locus with two alleles. It is assumed that double reduction does not occur.

The genotypic composition of some initial generation, say S0, is:

Genotype					
	aaaa	Aaaa	AAaa	AAAa	AAAA
	nulliplex	Simplex	Duplex	triplex	quadruplex
f	f_0	f_1	f_2	f_3	f_4

Its gene frequencies are

$$P = \frac{1}{4} f_1 + \frac{1}{2} f_2 + \frac{3}{4} f_3 + f_4$$

and

$$q = 1 - P$$

It is first verified that the gene frequencies remain constant from one generation to the next (such constancy is to be expected in the absence of selection). In order to do this, Table is used. This table presents, for each possible autotetraploid genotype, and according to the haplotype frequencies presented in Table, the genotypic composition of the line obtained by selfing. The allele frequencies in the parental population follow from Equation. Across the total of the lines obtained from this parental population the frequency of allele A is,

$$\frac{1}{4} \left(\frac{1}{2} f_1 + \frac{2}{9} f_2 \right) + \frac{1}{2} \left(\frac{1}{4} f_1 + \frac{1}{2} f_2 + \frac{1}{4} f_3 \right) + \frac{3}{4} \left(\frac{2}{9} f_2 + \frac{1}{2} f_3 \right)$$
$$+ \left(\frac{1}{36} f_2 + \frac{1}{4} f_3 + f_4 \right) = \frac{1}{4} f_1 + \frac{1}{2} f_2 + \frac{3}{4} f_3 + f_4$$

This is equal to the frequency in the parental population. The genotypic composition of S– will thus be:

Genotype:

	aaaa	Aaaa	AAaa	AAAa	AAAA
f	q	0	0	0	P

How fast do the frequencies of plants with a heterozygous genotype and of gametes with a heterozygous haplotype decrease with (continued) selfing?

genotype	f	aaaa	Aaaa	AAaa	AAAa	AAAA
aaaa	f_0	1	0	0	0	0
Aaaa	f_1	$\frac{1}{4}$	$\frac{1}{2}$	$\frac{1}{4}$	0	0
AAaa	f_2	$\frac{1}{36}$	$\frac{2}{9}$	$\frac{1}{2}$	$\frac{2}{9}$	$\frac{1}{36}$
AAAa	f_3	0	0	$\frac{1}{4}$	$\frac{1}{2}$	$\frac{1}{4}$
AAAA	f_4	0	0	0	0	1

In order to answer this question, first the decrease of $g1$, *i.e.* the frequency of gametes with haplotype Aa is considered and thereafter the decrease of fh. *i.e.* the frequency of heterozygous plants. From Table it can be derived that,

$$g1,t+1 = \frac{1}{2}f1,t + \frac{4}{6}f2,t + \frac{1}{2}f3,t$$

Thus similarly,

$$g_{1,t+2} = \frac{1}{2}f_{1,t+1}\frac{4}{6}f_{2,t+1} + \frac{1}{2}f_{3,t+1}$$

$$= \frac{1}{2}\left(\frac{1}{2}f_{1,t} + \frac{2}{9}f_{2,t}\right) + \frac{4}{6}\left(\frac{1}{4}f_{1,t} + \frac{1}{2}f_{2,t} + \frac{1}{4}f_{3,t}\right) + \frac{1}{2}\left(\frac{2}{9}f_{2,t} + \frac{1}{2}f_{3,t}\right)$$

$$= \frac{5}{12}f_{1,t} + \frac{5}{9}f_{2,t} + \frac{5}{12}f_{3,t}$$

$$= \frac{5}{6}g_{1,t+1}$$

This implies that each population obtained by selfing still produces 5 6 of the proportion of gametes with the Aa haplotype which was produced by the previous generation. Now the frequency of plants with a heterozygous genotype is considered. This frequency is designated by fh.

Thus,

$$f_{h,t} = f_{1,t} + f_{2,t} + f_{3,t}$$

$$f_{1,t+2} = \frac{1}{2}f_{1,t+1} + \frac{2}{9}f_{2,t+1}$$

$$f_{2,t+2} = \frac{1}{4}f_{1,t+1} + \frac{1}{2}f_{2,t+1} + \frac{1}{4}f_{3,t+1}$$

$$f_{3,t+2} = \frac{2}{9}f_{2,t+1} + \frac{1}{2}f_{3,t+1}$$

The decrease of f_h at (continued) selfing is described by:

$$f_{h,t+2} = \frac{3}{4}f_{1,t+1} + \frac{17}{18}f_{2,t+1} + \frac{3}{4}f_{3,t+1}$$

$$= f_{h,t+1} - \left(\frac{1}{4}f_{1,t+1} + \frac{1}{18}f_{2,t+2} + \frac{1}{4}f_{3,t+1}\right)$$

$$= f_{h,t+1} - \left[\frac{1}{4}\left(\frac{1}{2}f_{1,t} + \frac{2}{9}f_{2,t}\right) + \frac{1}{18}\left(\frac{1}{4}f_{1,t} + \frac{1}{2}f_{2,t} + \frac{1}{4}f_{3,t}\right) + \frac{1}{4}\left(\frac{2}{9}f_{2,t} + \frac{1}{2}f_{3,t}\right)\right]$$

$$= f_{h,t+1} - \frac{5}{36}(f_{1,t} + f_{2,t} + f_{3,t})$$

$$= f_{h,t+1} - \frac{5}{36}f_{h,t}$$

We consider the decrease of the frequency of heterozygous plants for an initial population consisting exclusively of duplex plants. The genotypic composition of S0 is then (0, 0, 1, 0, 0), with $f_{h,0} = 1$.

Generation	t	$f_{h,t}$	$\lambda_S = \dfrac{f_{h,t}}{f_{h,t-1}}$
S_0	0	1	
S_1	1	$\dfrac{17}{18} = 0.9444$	0.9444
S_2	2	$\dfrac{29}{36} = 0.8056$	0.8529
S_3	3	$\dfrac{437}{648} = 0.6744$	0.8372
S_4	4	$\dfrac{729}{1296} = 0.5625$	0.8341

Then to 2/9 + 1/2 + 2/9 = 17/18. Table presents the frequency of plants with a heterozygous genotype in successive generations, as calculated from Equation.

The frequency of heterozygous plants as a proportion of the frequency in the preceding generation, i.e.

$$\lambda_S = \frac{f_{h,t}}{f_{h,t-1}}$$

is also presented in Table. It appears that λS converges to a constant value, viz. to 5/6 = 0.8333. This implies, per round of reproduction by selfing, the same constant (relative) decrease in the frequency of heterozygous plants as derived from the frequency of heterozygous gametes. In this phase, reproduction by means of self-fertilization for n successive generations reduces $f_{h,t}$ to

$$f_{h,t+n} = \left(\frac{5}{6}\right)^n f_{h,t}$$

The frequency of heterozygous plants is halved if

$$\left(\frac{5}{6}\right)^n = 0.5, i.e. \ if$$

$$n = \frac{In(0.5)}{In(0.8333)} = 3.8$$

Starting with an initial population with genotypic composition (0, 0, 1, 0, 0) the decrease of the frequency of heterozygous plants is even less: in S4, $fh,4$ is still larger than 1/2. When comparing the decrease in the frequency of plants with a heterozygous genotype occurring at selfing of a diploid crop and such decrease at selfing of a diploid crop and such decrease at selfing of a tetraploid crop it is clear that the decrease is quite slow in the case of tetraploidy.

Continued FS-mating in a diploid crop gives a somewhat faster decrease in the frequency of heterozygous plants than continued selfing of a tetraploid crop. A more comprehensive treatment of population genetical effects of selfing in an autotetraploid population is given by Seyffert.

Chapter 4

Genetic Improvement of Plant

Introduction

Bamboos are among the economically most important plants world-wide. In Europe bamboos are used as ornamentals for gardens, but there is increasing interest for uses in ecological applications and as energy crops. Biotechnological techniques, including tissue culture, in vitro hybridisation, molecular markers and genetic transformation are crucial for the future of bamboo.

Micropropagation of bamboos has allowed developing a new type of ornamental bamboo that can be produced yearround with a high quality/price ratio and distributed far more widely than classically propagated ornamental bamboos. Molecular markers are used in quality control procedures.

Flowering of bamboos is still one of the greatest mysteries in botany, and breeding systems are non-existent. However, flowering can be induced reproducibly in tissue culture, both in seedlings and in adult bamboos, providing the only method for hybridisation.

The flowering structures that are used are pseudospikelets, morphological features unique to the subfamily of Bambusoideae. These special propagules can be used for propagation, long term storage, for hybridisation and for genetic transformation. While flowering can be induced, controlled and reversed in tissue culture, a more fundamental approach to unravel the mechanisms of flowering include studies of cell division patterns and profiles of volatile components.

Some applications of biotechnology for bamboo are presented, with emphasis on research strategies in a SME. Furthermore, all the techniques developed are of use not only in horticulture, but also in agriculture and forestry worldwide. Bamboo is an ornamental plant, mainly for garden use in Europe.

Besides, two different pathways for valorisation of biotechnological research have emerged:

- In the tropics bamboo is a very important plant, providing livelihood for over 500 million people and providing housing and shelter for over 1 billion people;
- "Bamboo for Europe", an EU- funded research project has allowed to develop the potential of bamboo as an agricultural plant in Europ with possible applications in wood industry and as energy crop.

So, any biotechnological developments for the improvement of ornamentals will also be useful on a much larger scale. To develop a commercially feasible process of bamboo propagation and genetic improvement three major components are important:

1. Fundamental research on bamboo physiology and genetics, including the development of tissue culture techniques for hybridisation of bamboo;
2. The development of axillary branching into a universal technique with a high clonal fidelity and very high efficiency;
3. Forward integration to optimise the added value of micropropagation.

Micropropagation of Bamboo

Research on Bamboo Micropropagation

For bamboo different propagation techniq ues are available, such as seed propagation, clump division, rhizome and culm cuttings. But these methods suffer from serious drawbacks for large or mass scale propagation. For mass scale propagation (> 500 000 plants per year) classical techniques are largely insufficient and inefficient, and tissue culture is the only viable method. Indeed, the order of magnitude of the demand for bamboo planting materials indicates that micropropagation will inevitably be necessary for mass scale propagation.

By now a large number of papers on micropropagation (the use of tissue culture for propagation only) of bamboos have been published, original papers as well as reviews and some in which tissue culture is described in a more general aspect. Many researches have focussed on somatic embryogenesis of seedlings of tropical bamboos. An inventory shows that at least 21 labs in South-east Asia were involved in bamboo tissue culture, mainly in India. Major research focused also on the clonal propagation of elite genotypes, either juvenile or adult. The number of papers about this subject however is much less, and this is solely due to lack of success. Indeed, technically the propagation of adult plants via axillary branching is much more difficult than with seedlings of tropical bamboos.

For tissue culture of bamboo the use of starting material (seeds or adult plants) and the choice of the propagation method are crucial. The two major advantages of using seedlings are that seedlings establish a new generation, and that the techno logy is easier.

But the disadvantages are considerable:

* Insufficient or no knowledge of genetic background,
* Restricted availability of seeds for most species and rapid loss of germination capacity,
* Comparison of in vitro to in vivo performance has not been thoroughly evaluated.

In addition there is a huge variability in responsiveness in tissue culture.

When using adult bamboos main problems are:

* Endogenous contamination,
* Hyperhydricity and instability of multiplication rates,
* Many problems with rooting also in bamboos that root readily in nature.

Rooting percentages for adult bamboos ranged from very low percentages of 10 per cent for Bambusa vulgaris to 73 per cent for adult Dendrocalamus longispathus. A rooting percentage of 77 per cent was obtained for adult Dendrocalamus giganteus in 3 or 4 weeks. But, while 77 per cent of success is good on 500 plants in a laboratory experiment it still represents a loss of 33000 when we transplant 100 000 plants. Low rooting frequencies are the major bottleneck to developing commercially viable protocols. The combination of photomixotrophic in vitro multiplication and photoautotrophic in vitro rooting stages resulted in improved transplanting success. Improvement of rooting percentages and transplanting has been achieved in various commercial laboratories.

Commercial Micropropagation of Bamboo

Micropropagation via tissue culture thus attracted a lot of attention but the translation and transformation of these expectations into commercially viable propagation systems has been beset with a number of problems that were either technological, or related to marketing. At present only a very limited number of laboratories specialise in the production of bamboos at a commercial scale. Oprins Plant, Belgium, currently has over 60 genotypes in culture initiated from adult genotypes of which about 30 on a commercial scale.

Species and varieties of Arundinaria, Chimonobambusa, Fargesia, Phyllostachys, Pleioblastus, Sasa, Sasaella, Semiarundinaria, Shibataea and Yushania (temperate bamboos) and Bambusa, Dendrocalamus, Dinochloa, and Thyrsostachys (tropicals) are produced commercially. Rooting percentages of most species are between 85 and 100 per cent, depending on species and season of transplanting (unpublished results). These bamboos are produced either as ornamental, or for tropical forestry. Other laboratories focusing on bamboo tissue culture include West Wind Technology, USA, and Bamboo World, Australia.

Several other laboratories such as Piccoplant, Germany, and Microflor, Belgium, produce a more limited number of bamboo species as part of their tissue culture operation. Other laboratories initially involved in tissue culture of bamboos however, no longer produce bamboo commercially, for example TERI, India and Thai Orchids Lab, Thailand. The main problem is that micropropagation of bamboos also has to deal with the development of new markets and market opportunities. Logistics and supply chains such as in ornamental horticulture where tissue culture labs are separate entities seem not to work. Instead the laboratories have to develop their own markets, marketing and sales for bamboo plants. This strategy of forward integration needs to go beyond the classical scheme of tissue culture.

The tissue culture phase terminates at Stage III, but in regard to marketing of plants one can also distinguish subsequent stages:

- Stage IV, the transplantation stage with the end product being rooted plantlets in trays,
- Stage V, the production of liners, either for production of saleable plants or for use as micromotherplant,
- Stage VI, the production of saleable plants.

This distinction is important if the complete chain of production is integrated in a single company, since this determines the added values. In our case the laboratory produces ornamental plants exclusively for Oprins Plant's own nurseries. A balanced production and logistic chain, where bamboos are produced in Belgium and grown in the nurseries in Spain and France, has allowed to produce high quality ornamental bamboos year round at prices which are 40-60 per cent below the current market prices for bamboo. These plants are currently being sold under the brand name BambooSelectTM, which is an additional tool in marketing. In an agricultural perspective it is now possible to produce the same genotypes for large scale plantations in Europe at prices comparable to other energy crops.

Monitoring Genetic Stability in Bamboo Tissue Culture

AFLP Markers in Quality Control of Micropropagation

Somatic mutations in bamboo are highly valued in horticulture. In various bamboos somatic mutations are known that alter the stem colour or stem shape, such as Bambusa ventricosa with inflated internodes. But not all are stable in tissue culture.

The bulbous internodes of Bambusa ventricosa were lost after tissue culture and variegation of stems and leaves disappears very often in tissue culture. Tissue culture has been associated with molecular aberrations in general (including also chromosomal defects), and phenotypic defects may show up only some years later. Therefore it is important to devise methods of quality control at the tissue culture stages to minimise the potential danger of such defects, for exa mple when plants of Stage VI are sold in large quantities.

Generally it is very difficult to find precise defects, especially when somatic mutations are possibly linked to transposon activity. In a first approach AFLPTM markers are used to assess the stability and clonal fidelity of tissue cultured bamboos.

In one experiment Phyllostachys species and cultivars were collected:

- From the garden of Dr. Jacques Van Dooren, Kumtich, reference collection of bamboo for Belgium
- From the tissue culture lab at Oprins Plant
- from plants transplanted in the greenhouse.

Using this set up, it should be possible to detect if genotypes in the collection and in propagation have the correct names, and whether the plants cultured are subject to considerable clonal variation.

The most bamboos in tissue culture and transplanted in the greenhouse group together with the collection plants. The exception is Phyllostachys vivax from the laboratory that turns out to belong to the nigra group, while the collection plant of P. vivax does not group at all with the nigra's.

This is a clear case of mistaken identity, due to mislabelling of the motherplants obtained from China. If one observes the dendrograms it is seen tha t all P. aurea cultivars, all P. nigra cultivars, and all P. aureosulcata cultivars group together. They are, according to this test, 100 per cent identical. However, they differ in stem colour, leaf colour, habit etc, and these differences can be observed readily in the field. This also means that the method used is not sensitive enough to detect those differences. But it clearly shows that AFLP is a useful method when used as part of quality control in the tissue culture process, to assess correct clonal identity. This can then be used towards customers. AFLP markers are now used routinely besides other measures to ensure clonal fidelity. These other measures include the use of axillary branching as propagation method and annual initiation of new cultures, so that the maximum number of subcultures is 12 to 14.

Molecular Basis of Somatic Mutations in Bamboo

Based on phenotype and mode of action somatic mutations involved in stem colour of bamboo are possibly caused by transposon activity. The study of transposon activity and their possible relation to certain mutations is very difficult. In bamboo no breeding systems are available to detect transposon activity through classical breeding. Molecular approaches may provide some shortcuts, but to link the presence and activity of transposons to certain phenotypic markers is again a difficult task.

A first approach to identify the molecular basis of these somatic mutations was to use primers based on sequence information from the 4.5 kb Ac9 transposon from maize. Several primer combinations were used and several copies with considerable homology to the original Ac9 transposons were present, which eventually turned out to be a useful and relatively cheap method to identify bamboos, at the species level.

Primers designed to detect Activator-like transposons in Petunia could also be used to detect fragments in bamboo. The same fragment was detected in the three bamboos tested, namely Bambusa vulgaris, Sasa veitchii and Phyllostachys edulis. The fragment of Bambusa vulgaris (hATbv1) was sequenced and shown to be homologous to members of the hAT superfamily of transposons, to which also Ac (corn), Tam (Antirrhinum majus) and hobo (Drosophila) belong. The homology was as high as 60 per cent in some regions, which conforms also to earlier findings of Ac-like sequences in Bambusa multiplex. It is very likely that many other transposons are present in the bamboo genome as well.

A next step is to use these AFLP-primers in combination with methylation sensitive AFLP, in which methylation sensitive restriction enzymes are used to cleave sites within the bamboo transposon. The relation between methylation patterns and transposon activities is well established and the activation of transposons during tissue culture is also a well documented phenomenon.

The use of genotypes that have been cultured for six, eighteen, twenty-four and thirty-six months should allow to assess if long term tissue culture has any influence on methylation in general and on methylation patterns of transposons. This research is currently ongoing.

Flowering of Bamboo in Tissue Culture

Because of the peculiar flo wering habits in bamboo it has been almost impossible to breed for superior traits in woody bamboos. The first reports on tissue culture flowering of bamboos caused great excitement. It opened up the possibility of controlled flowering that can be used for breeding of bamboo. Since then in vitro flowering has been observed in many types of bamboos, both in seedlings and mature bamboos. More than 10 years after the first report on in vitro flowering in bamboo some positive results have been obtained, but practical and commercially exploitable results have not been reported yet. The most promising potential is for hybridisation under controlled conditions. But many hurdles still need to be taken before the methods really become applicable at agricultural scales.

Recently a new research project was initiated in our group to improve the induction of flowering and the quality of flowering parts, in order to be able to hybridise bamboos in the future. While this project aims mainly at developing techniques, it also includes various approaches on a more fundamental level. The use of pseudospikelets is central in this research.

Pseudospikelets occur only in bamboos, not in other grasses. In grasses a spikelet bears two empty glumes, but the glumes of pseudospikelets in bamboos subtend dormant buds that can develop into new pseudospikelets. The proximal parts of the pseudospikelets do not develop into flowers but in tissue culture conditions keep on multiplying indefinitely, allowing to establish monocultures of pseudospikelets. Flowers are formed in the distal parts of individual pseudospikelets and do not develop to anthesis under normal tissue culture conditions.

They can however, be forced to develop fully under more adverse conditions. Cultural conditions can control flowering in vitro to a large extent. Factors used for the induction of pseudospikelets are mainly cytokinins. However, cytokinins are involved in various flowering processes and more fundamental research is urgently needed. Unfortunately, despite the huge potential of tissue culture flowering of bamboo, to our knowledge only one other laboratory is actively engaged in this field of research.

Despite the presence of flowers in the axils of pseudospikelets, dormant buds can develop into vegetative shoots. Pseudospikelets can also be regarded as highly contracted axes, in which leaves are reduced to glumes. Instead of propagules and plantlets of 2 to 5 cm in tissue culture, pseudospikelets have lengths between 0.25 and 1.5 cm. This is a miniaturisation of the plants and allows a good opportunity to use these pseudospikelets for propagation of bamboos.

With these smaller clumps, operators can cut and transfer much easier, while the mode of propagation is still via axillary branching, thus the method with lesser risk for somaclonal variation. Since leaf area is much reduced, quality problems in the propagation stage are much less important. This allows for a considerable cost reduction in micropropagation, since operator costs represent 75- 80 per cent of cost price of tissue culture plants.

In tissue culture systems for propagation flowering of bamboo represents considerable risks since many bamboos flower monocarpically in nature and die completely after flowering. It is very important to be able to control flowering and reversion of flowering of bamboo. Therefore the influence of tissue culture flowering on vegetative growth post vitro has been studied extensively in three independent experiments (unpublished results).

- Plants of Bambusa tuldoides regenerated from pseudospikelets grow as vegetative plants after hardening;
- In large scale vegetative propagation only two flowering plants of Phyllostachys have been observed after transplanting about 500000 plants,

- Tests with Sasa palmata have shown that even when flowering plants are transplanted in the greenhouse, these revert to the vegetative state within a few weeks.

These tests have shown that tissue culture technology does not lead to monocarpic flowering of saleable plants. Using pseudospikelets as propagules we have now a system which can be used for propagation and for the induction of flowering and flowering structures. The same propagules can also be used for long term storage and for genetic transformation.

Tissue Culture Techniques

Fungi and bacteria grow on the plant surface and contaminate the culture media when they are not adequately eliminated. The process of explant introduction into in vitro conditions depends mainly on the disinfecting phase. Problems in handling produce contamination of the in vitro plant and so, a similar disinfection process to that of the in vitro introduction technique is recommended.

However, it is possible to use antibiotics included in the medium, but just temporarily, while the plantlets are growing. Among the bacterial contaminants are Bacillus sp., Erwin/a sp., Pseudomonas sp., etc. In the case of systemic infections, the use of disinfectants is not effective, because the pathogens are located in the vascular system. In this case, it is recommended to use specific antibiotics and meristems or bud cuttings.

On the other hand, the process of in vitro introduction involves the use of explants in different physiological stages, as dormant buds, which require growth regulators for example, Gibberellic acid) to stimulate and accelerate bud growth. During the process of in vitro introduction there is a close connection between the greenhouse and the laboratory; that is why the necessary precautions should be taken to avoid the entrance of contaminants into the laboratory.

- In the greenhouse
  - Take cuttings from the mother plant, or buds from greenhouse tuber sprouts.
- In the laboratory
  - Immerse the cuttings for 10 minutes in a beaker conveniently labeled with the accession name, and containing a solution of 0.5 per cent acaricide with three drops/I of Tween 20.

 - Throw away the acaricide solution and rinse the nodes with tap water
 - Prepare the laminar flow chamber.
 - Immerse the cuttings in 700/o alcohol for 30 seconds and take the beakers immediately to the laminar flow chamber
 • In the laminar flow chamber
   - Eliminate the alcohol and replace it with a 2.50/o solution of calcium hypochlorite. Keep the cuttings immersed during 15 minutes.
   - Throw away the hypochlorite, and rinse the cuttings three times with sterile water
   - Keep the nodes immersed in sterile water until bud extraction.
   - Dissect the buds on a sterile plate and place them in the culture medium.

Micropropagation

In vitro plantlets, which are free of pathogens, are used as initial material for potato and sweetpotato seed programmes. The methods used in these micropropagation programmes mainly depend on their production volume and the available infrastructure.

In the case of potato micropropagation, the basic methods have been already described. They have been verified in many institutions and they are based on the rapid growth of individual node cuttings, or stems with multiple node cuttings. Afterwards, the basic micropropagation methods are described.

Node micropropagation

This method is based on the principle that the node of an in vitro plantlet placed in an appropriate culture medium will induce the development of the axillary bud, resulting in a new in vitro plantlet. This type of propagation promotes the development of a pre-existent morphological structure. The nutritional and hormonal condition of the medium breaks the dormancy of the axillary bud and promotes its rapid development.

Callus formation and plant regeneration must be avoided because they tend to affect the genetic stability of the genotype. Under room-controlled conditions micropropagation is fast. Each node planted in a propagation medium will produce a plantlet which will occupy the full length of the test tube, after approximately four weeks for potato, and six weeks for sweetpotato.

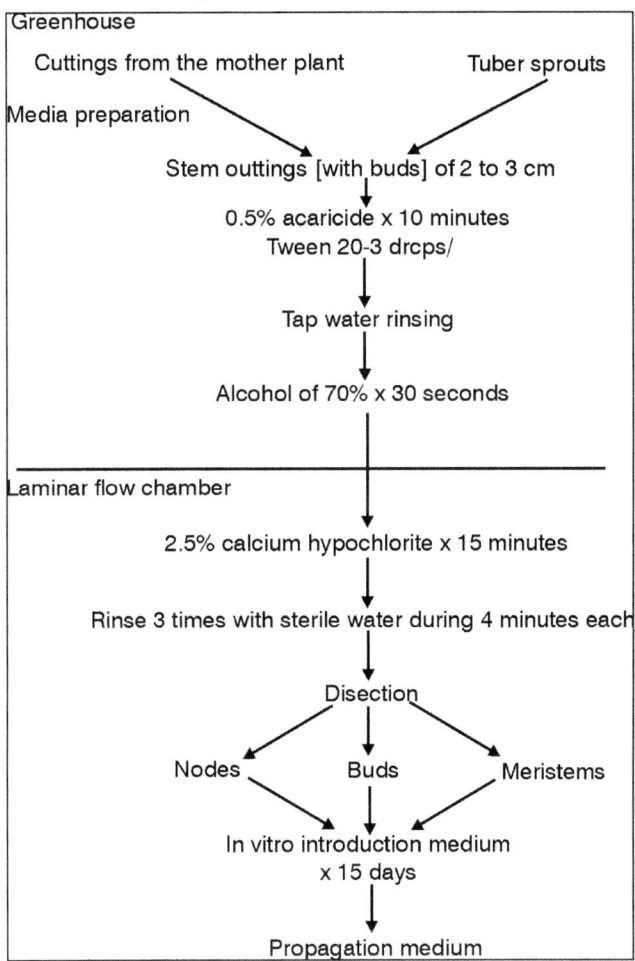

Figure 1: Procedure for in vitro Introducction

The resultant in vitro plantlets may be transplanted to in vitro conditions in small pots in the greenhouse.

Micropropagation by Node Cuttings in a Liquid Medium

This technique is applied both with potato and sweetpotato to produce a large number of nodes rapidly. Stem cuttings with 5 to 8 nodes are prepared by removing both the apex and the root of the in vitro plant to be propagated. The stems are placed in the corresponding propagation liquid medium. It is also possible to use isolated nodes: the nodes will sprout and new plantlets will develop over a period of 3 to 4 weeks.

Micropropagation Procedure

- Sterilize petri dishes (placed in paper bags or comets) and prepare the laminar flow chamber by disinfecting the internal surfaces with alcohol.

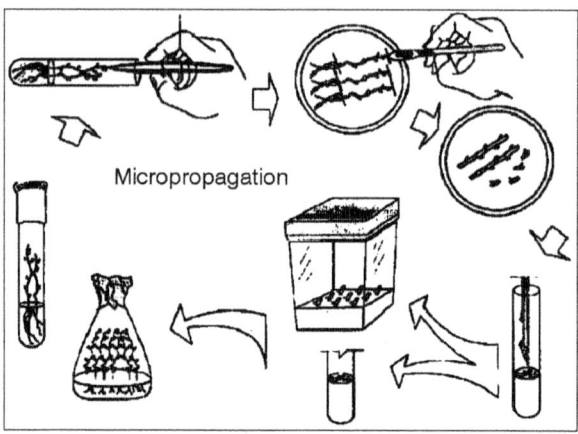

Figure 2: Potato Micropropagation Process Scheme

Sterilize the tools with an instrument sterilizer and place them on a sterile dish.

- Open the tube, take off the plantlet and place it on a petri dish with the help of forceps.
- Remove the leaves and cut the nodes.
- Open a tube containing fresh sterile medium and place a node inside, trying to plunge it slightly into the medium with the bud up. Close the tube.
- Seal the tube with a gas-permeable plastic tape (parafilm or saram wrap) and label it correctly. It is recommended to place two explants in 16×125 mm tubes, three in 18×150 mm tubes, five in 25×150 mm tubes, and 20-30 in magenta vessels.

Common Problems in Micropropagation

Some problems may appear in tissue culture according to the crop or variety. To solve them, it is necessary to apply one or several preventing/solving methods such as:

Phenolization: The explants frequently become brown or blackish shortly after isolation. When this occurs, growth is inhibited and the tissue generally dies. The young tissues are less susceptible to darkening than the more mature ones.

Prevention: The tissue darkening–mainly that of the recently isolated explants and that of the medium–may be generally prevented by:

- Removing the phenolic compounds produced by dispersion. Absorption by means of activated carbon. Absorption by polyvinilpyrolidone (PVP).

- Modifying the redox potential. Reducing agents: ascorbic acid, citric acid, L-cisteine HCL, ditriotreitol, glutation and mercaptoethanol Less availability of oxygen: stationary liquid media.

- Inactivating the phenolase enzymes. Chelating agents: NaFeEDTA, EDTA, diethyldithiocarbamate, dimethyl-dithiocarbamate

- Reducing the phenolasic activity and the availability of substrate. Low pH Darkness

Absence of Rooting: The explants can naturally form roots during propagation, without an additional rooting stage, as with the potato. However, some wild potato species may show root production deficiency. Rooting may be induced by incorporating auxins, such as IAA, NAA, and IBA, or activated carbon to the culture medium.

Potato in Vitro Tuberization

Most of the potato microtubers are used in seed programmes in Europe, where large amounts (hundred of thousands) of pre-basic seed of a few varieties are produced. By means of this technique, microtubers are produced and stored, and it is possible to store thousands of them in a small area (in humid containers at 4°C) for long periods of time. The microtuber induction is produced through a stress effect by the CCC (chlorocholine chloride), BAP (B-benzylaminopurine) and Sucrose, which under darkness will produce from 3 to 4 microtubers per plant, according to the variety.

In the beginning, microtubers were used as an alternative for germplasm distribution and in vitro conservation. Curing the plantlet in vitro multiplication process, usually there is a multiplication of higher amounts to those required in the greenhouse and so, after the transference, some magentas are left over. These plantlets may be used for microtuber induction.

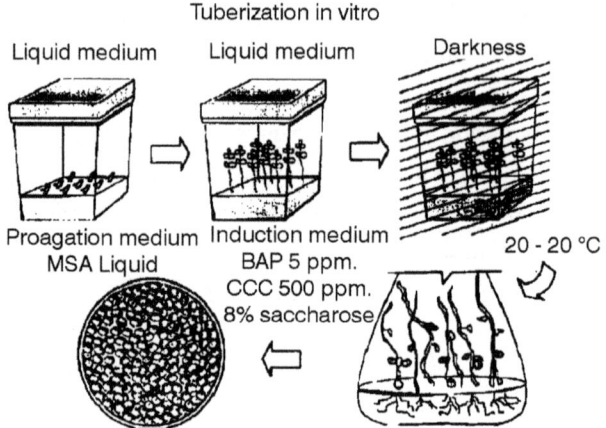

Figure 3: Poato in Vitro Tuberization Process Scheme

In accordance with the indicated procedures, the induction medium is added to the magentas and then they are placed in the dark room. After three months the microtubers will be produced. These may be harvested and transferred to sterile containers (at 4°C), where they can be maintained up to 10 months, or used in the next campaign instead of in vitro plantlets.

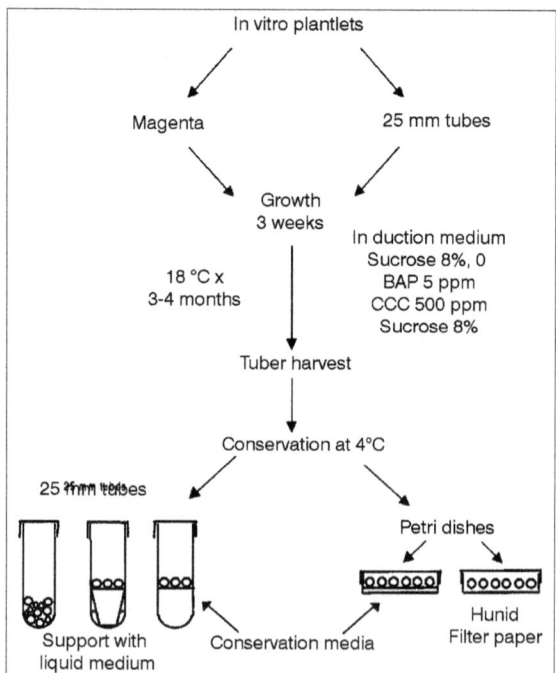

Figure 4: Potato in Vitro Tuberization Process Scheme

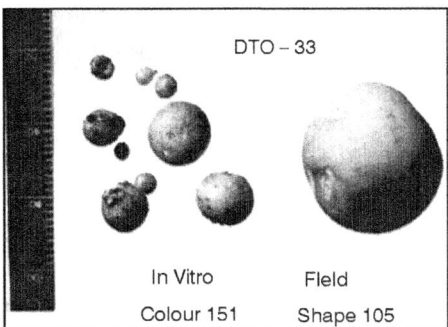

Figure 5: Comparison, in Size and Shape of in Vitro Microtubers with a Normal one from the Field

These microtubers can also be used as a reserve, in case in vitro material is contaminated or dies, because of temperature or handling effects. The high number and bigger size of the microtubers increases the success rates for the transfer to the greenhouse.

Medium Term Germplasm Maintenance

The in vitro germplasm maintenance under normal growth conditions requires a series of transfers of the plantlets into a fresh medium. This leads to a consumption of time, increases the possibility of losses because of material contamination during successive sub-cultures, causes a loss of material by human error or failure of some equipment, and demands more labour A way to avoid these problems is through limitation, restriction or inhibition of growth. This approach consists of growth speed reduction by modifying the physical or chemical conditions of the culture, and it is effective for a short or medium term period.

In Vitro Conservation Methods: These consist in maintaining the cultures (buds, plantlets derived from nodes or directly from meristems) under physical (environmental factors] or chemical (culture medium composition) stressed conditions that make it possible to extend, as much as possible, the interval of transference into the fresh media, without affecting the viability of the cultures. The methods to reduce the in vitro cultivated plant growth include the reduction of temperature and light during storage, the incorporation of growth retardants in the medium, and the induction of osmotic stress in the medium, or a combination of all these.

Temperature: Temperature reduction has been the most used way to curb culture growth. Most of the in vitro cultures are maintained at temperatures between 12 and 200C: at lower temperatures, the growth rate decreases but the reduction depends on the species.

Nutrients Concentration: The reduction of the carbohydrate concentration and the nitrogenous components of the nutritive medium may affect the growth rates. In addition, the continuous absorption of these nutrients during plantet growth will bring a nutritional defficiency, which could produce a premature death of the plants.

Use of Growth Regulators: The abscisic acid (ABA), phosphon-D, maleic hydrazide, and succinic acid are some of the most frequently used growth regulators.

Osmotic Concentration: Growth limitation caused by osmotic concentration is due to the reduction of the water and nutrients absorbed from the medium. For example, at high concentrations, sucrose acts osmotically and it is highly metabolized. No metabolized osmotics, such as Manitol and Sorbitol, are possibly more efective than sucrose in culture growth limitation.

Evaluation of in Vitro Preserved Material: To evaluate material under these conditions, some important facts about in vitro maintenance such as genetic viability and stability should be considered. The viability evaluation of the in vitro cultures must be systematic. In slow growth conditions, when the sub-culture or transfer period extends during months or years, the frequency of the culture evaluation increases. The most important characteristics to be evaluated in the low-growth cultivars storage of apical buds are: contamination, leaf senescence, the number of green sprouts, the number of viable nodes in relation to the stem length, the presence or not of roots, and callus formation.

Maintenance of Accessions in Seed Programme

In a seed programme, a group of accessions free of virus is maintained for pre-basic seed production: however, most of them are not propagated in the greenhouse so they are maintained in vitro to be used in the future. The continuous propagation of the in vitro plantlets damages the material, mainly if we consider that environmental conditions are not adequate.

The alternative is to maintain the accessions in conservation media. Each accession must be maintained in test tubes with five replications to avoid possible losses. The maintenance conditions have been tested in CIP's potato collection that consists of more than 5,000 accessions, by means of which it is possible to assure the normal recovery of the material after using stress producers. The plantlets used in each campaign must come from the maintenance phase. Then, they will be propagated in normal media where their growth will be reestablished, and the elected procedures for plantlet multiplication in the greenhouse will continue.Each season must be initiated with this material to start off with strong plants. In addition, the plantlets taken for multiplication must be replaced, trying to maintain 5 tubes per accession in the conservation media.

The sub-cultures in the conservation medium are carried out approximately every one or two years. The conservation medium renovation must pass through a previous sub-culture in the propagation media to rejuvenate the explants.

Virus Eradication Through Meristems Culture And Thermotherapy

If a healthy plant is sown in the field, it is exposed to infections caused by pathogens as nematodes, fungi, bacteria, phytoplasms, virus and viroids, which have a negative effect on yield, and in some cases may kill the plants. However, not all the plant cells may become infected. A group of cells, which are in a continuous non-differentiated multiplication, are virus-free: the meristem.

The in vitro meristems culture, together with growth at high temperatures produces potato plantlets free from viruses in more than 90 per cent of planted meristems. This routine method was established in the International Potato Centre to obtain virus-free plantlets for national and international distribution.

The in vitro maintenance of virus-free plants provides the possibility to maintain all the time a bank of healthy and more vigourous plants, and with a more accelerated growth, than the infected ones. In a seed programme, it is essential to initiate this work with virus-free plantlets since this will affect the seed quality and the yield as well. The virus cleaning procedure is too long and expensive: that is why CIP, through its germplasm distribution programme, has provided a list of the principal virus-free potato varieties to all users.

Procedure

- Approximately 18 to 20 plantlets (virus-infected) are propagated in magentas.
- After a growth period of 20 to 25 days, or when the plantlets are 4-5 cm high, they are placed in the thermotherapy chamber. The growth conditions are: 16 hours of light 34°C; 8 hours of darkness 32°C

- The magentas are maintained in the thermotherapy chamber for one month.
- Afterwards, the magentas are taken out of the chamber, and the outside is cleaned with 980/o alcohol. Then they are introduced to the culture room.
- Meristems are obtained as follows:

Cut the apical portion and remove the leaves that cover the meristem (approximately 3 to 4 leaves); the meristem is observed with a prominent leaf primordium. Remove the meristem with part of the leaf primordium; cut only the translucent portion. Use a new knife.

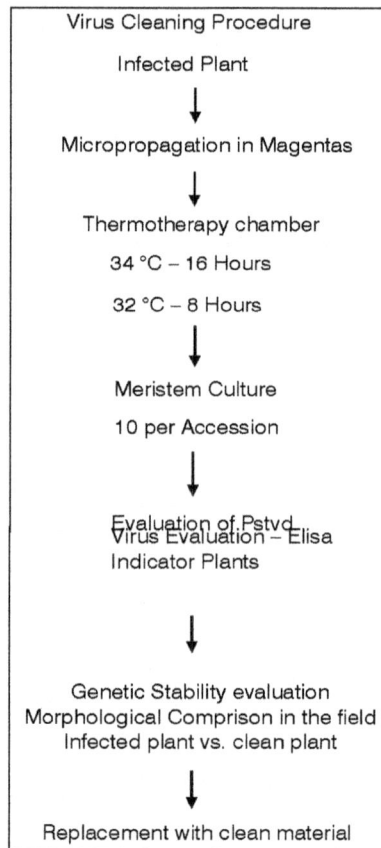

Place the meristem in the culture medium. Be sure that the meristem is in the tube.

- Evaluate meristem growth and transfer them to fresh media if necessary.

- Each meristem that originates a plant is called <'line», which will be labeled according to the accession it belongs to. For example, Yellow Line 1, Yellow Line 2, Yellow Line 3, etc.
- Five tubes with several plants are propagated to evaluate the potato virus PSTVd, PVT, to determine the host range, the morphologic evaluation, and the in vitro maintenance.
- The results of each evaluation are obtained, and the infected material is replaced with clean material.

Temperature: The growth of the plantlets at temperatures higher than 20°C is accelerated (Do not use temperatures higher than 30°C).

Cultured Cells and Tissues

The large scale commercial propagation of plant material based on plant tissue culture was pioneered in the USA. During the last thirty years, tissue culture-based plant propagation has emerged as one of the leading global agro-technologies. Between 1986 and 1993, the worldwide production of tissue cultured plants increased 50 per cent. In 1993, the production was 663 million plants. By 1997, production had risen to 800 million plants. During 1990–1994, the micropropagation industry declined in Europe, mainly due to production shifting to developing countries, but since then because of the demand for high quality and number, production in European countries has increased.

Since 1995, production has increased by 14 per cent in Asian countries, mainly due to the market entry of China, while the increase in South and Central America was from production in Cuba. More recently, some companies from Israel, the USA and UK have shifted their production requirements to Costa Rica and India.

Tissue cultured plants have as yet to reach many growers and farmers in the developing countries. The primary advantage of micropropagation is the rapid production of high quality, disease-free and uniform planting material. The plants can be multiplied under a controlled environment, anywhere, irrespective of the season and weather, on a year-round basis. Production of high quality and healthy planting material of ornamentals, and forest and fruit trees, propagated from vegetative parts, has created new opportunities in global trading for producers, farmers, and nursery owners, and for rural employment. Micropropagation technology is more expensive than the conventional methods of plant propagation, and requires several types of skills. It is a capital-intensive industry, and in some cases the unit cost per plant becomes unaffordable. The major reasons are cost of production and know-how. During the early years of the technology, there were difficulties in selling tissue culture products because the conventional planting material was much cheaper.

Now this problem has been addressed by inventing reliable and cost effective tissue culture methods without compromising on quality. This requires a constant monitoring of the input costs of chemicals, media, energy, labour and capital. In the industrialized countries, labour is the main factor that contributes to the high cost of production of tissue-cultured plants.

To reduce such costs, some steps can be partially mechanized, e.g. use of a peristaltic pump for medium dispensing, and of dishwashers for cleaning containers. In the less developed countries of Africa, Asia, and Latin America, where labour is relatively cheaper, consumables such as media, culture containers, and electricity make a comparatively greater contribution to production costs.

For example, the cost of medium preparation (chemicals, energy and labour) can account for 30–35 per cent of the micropropagated plant production. However, automated production processes based on pre-sterilized membrane capsules, bioreactors, mechanized explant transfer, and container sealing are not commercially viable propositions in many developing countries. Therefore, low cost alternatives are needed to reduce production cost of tissue-cultured plants.

Many of the low cost technology options described in this publication can be incorporated in various steps of plant micropropagation. The occasional tissue culturegenerated variants (somaclones) and rare spontaneous bud mutants as well as those obtained from induced mutations can also be propagated by deployment of low cost techniques described. Micropropagation, in combination with radiation-induced mutations, speeds up the recovery, multiplication and release of improved varieties in vegetatively propagated plants. Hence, low cost technology will be of great value for large scale plant multiplication of mutants of many fruits, shrubs, flowers and forest trees that are conventionally vegetatively propagated.

Plant Tissue Culture

Plant tissue culture refers to growing and multiplication of cells, tissues and organs of plants on defined solid or liquid media under aseptic and controlled environment. The commercial technology is primarily based on micropro-pagation, in which rapid proliferation is achieved from tiny stem cuttings, axillary buds, and to a limited extent from somatic embryos, cell clumps in suspension cultures and bioreactors. The cultured cells and tissue can take several pathways. The pathways that lead to the production of true-to-type plants in large numbers are the preferred ones for commercial multiplication. The process of micropropagation is usually divided into several stages *i.e.*, prepropagation, initiation of explants, subculture of explants for proliferation, shooting and rooting, and hardening. These stages are universally applicable in large-scale multiplication of plants. The delivery of hardened small micropropagated plants to growers and market also requires extra care.

Plant tissue culture refers to growing and multiplication of cells, tissues and organs on defined solid or liquid media under aseptic and controlled environment. Plant tissue culture technology is being widely used for large-scale plant multiplication. The commercial technology is primarily based on micropropagation, in which rapid proliferation is achieved from tiny stem cuttings, axillary buds, and to a limited extent from somatic embryos, cell clumps in suspension cultures and bioreactors.

Explant Source

Plant tissue cultures are initiated from tiny pieces, called explants, taken from any part of a plant. Practically all parts of a plant have been used successfully as a source of explants. In practice, the "explant" is removed surgically, surface sterilized and placed on a nutrient medium to initiate the mother culture, that is multiplied repeatedly by subculture. The following plant parts are extensively used in commercial micropropagation. Shoot-tip and meristem-tip culture: Shoots develop from a small group of cells known as shoot apical meristem. The apical meristem maintains itself, gives rise to new tissues and organs, and communicates signals to the rest of the plant. Shoot-tips and meristem-tips are perhaps the most popular source of explants to initiate tissue cultures.

The shoot apex explant measures between 100 to 500 µ m and includes the apical meristem with 1 to 3 leaf primordia. The apical meristem of a shoot is the portion lying distal to the youngest leaf primordium, and is ca.100 µ??m in diameter and 250µ?m in length with 800-1200 cells. In practice, shoot-tip explants between 100 to

1000 μ m are cultured to free plants from viruses. Even explants larger than 1000 μ m have been frequently used. The term "meristem-tip culture" has been suggested to distinguish the large explants from those used in conventional propagation.

Nodal or axillary bud culture: This consists of a piece of stem with axillary bud culture with or without a portion of shoot. When only the axillary bud is taken, it is designated as "axillary bud" culture. Floral meristem and bud culture: Such explants are not commonly used in commercial propagation, but floral meristems and buds can generate complete plants. Other sources of explants: In some plants, leaf discs, intercalary meristems from nodes, small pieces of stems, immature zygotic embryos and nucellus have also been used as explants to initiate cultures.

Cell suspension and callus cultures: Plant parts such as leaf discs, intercalary meristems,-stem-pieces, immature embryos, anthers, pollen, microspores and ovules have been cultured to initiate callus. A callus is a mass of unorganized cells, which in many cases, upon transfer to suitable medium, is capable of giving rise to shoot-buds and somatic embryos, which then form complete plants. Such calli on culture in liquid media on shakers are used for initiating cell suspensions. Liquid suspension cultures maintained on mechanical shakers achieve fast and excellent multiplication rates. However, in commercial micropropagation, calli are cultured mostly in bottles and flasks kept on semi-solid or liquid media. To a limited extent, bioreactors have become popular for somatic embryogenic cultures. It is considered that some day robotics could be adapted to bioreactorbased micropropagation.

Pathways of Cultured Cells and Tissues

The cultured cells and tissue can take several pathways to produce a complete plant. Among these, the pathways that lead to the production of true-to-type plants in large numbers are the popular and preferred ones for commercial multiplication. The following terms have been used to describe various pathways of cells and tissue in culture.

Regeneration and Organogenesis

In this pathway, groups of cells of the apical meristem in the shoot apex, axillary buds, root tips, and floral buds are stimulated to differentiate and grow into shoots and ultimately into complete plants. In many cases, the axillary buds formed in the culture undergo repetitive proliferation, and produce large number of tiny plants.

The plants are then separated from each other and rooted either in the next stages of micropropagation or in vivo (in trays, small pots or beds in glasshouse or plastic tunnel under relatively high humidity). The explants cultured on relatively high amounts of auxin (*e.g.* (2,4-D, 2,4-dichlorophenoxyacetic acid) form an unorganized mass of cells, called callus. The callus can be further sub-cultured and multiplied.

The callus shaken in a liquid medium produces cell suspension, which can be subcultured and multiplied into more liquid cultures. The cell suspensions form cell clumps, which eventually form calli and give rise to plants through organogenesis or somatic embryogenesis. In some cases, explants *e.g.* leaf-discs and epidermal tissue can also generate plants by direct organogenesis and somatic embryogenesis without intervening callus formation, *e.g.* in orchardgrass. Dactylis glomerata L.. In organogenesis the cultured plant cells and cell clumps (callus) and mature differentiated cells (microspores, ovules) and tissues (leaf discs, inter-nodal segments) are induced to differentiate into complete plants to form shoot buds and eventually shoots, and rooted to form complete plants.

Somatic Embryogenesis

In this pathway, cells or callus cultures on solid media or in suspension cultures form embryo-like structures called somatic embryos, which on germination produce complete plants. The primary somatic embryos are also capable of producing more embryos through secondary somatic embryogenesis. Although, somatic embryogenesis has been demonstrated in a very large number of plants and trees, the use of somatic embryos in large-scale commercial production has been restricted to only a few plants, such as carrot, date palm, and a few forest trees.

Somatic embryos are produced as adventitious structures directly on explants of zygotic embryos, from callus and suspension cultures. Somatic embryos and synthetic seeds (embryos encapsulated in artificial endosperm) hold potential for large-scale clonal propagation of superior genotypes of heterogeneous plants.

They have also been used in commercial plant production and for the multiplication of parental genotypes in large-scale hybrid seed production. In many species, somatic embryos are morphologically similar to the zygotic embryos, although some biochemical, physiological and anatomical differences have been documented. The synthetic auxin, 2,4-D is commonly used for embryo induction. In many angiosperms, *e.g.*, carrot and alfalfa, subculture of cells from 2,4-D containing medium to auxin-free medium is sufficient to induce somatic embryogenesis. The process can be enhanced with the application of osmotic stress, manipulation of medium nutrients, and reducing humidity. Selection of embryogenic cell lines has also been successfully used. For example, selection for unique morphotypes in grapevine cultures allows production of high quality embryos with predicable frequency.

A major problem in large-scale production of somatic embryos is culture synchronization. This is achieved through selecting cells or pre-embryonic cell clusters of certain size, and manipulation of light and temperature, temporary starvation or by adding cell cycle synchronizing chemicals to the medium. Cytokinins seem to play a key role in cell cycle synchronization and embryo induction, proliferation and differentiation. Abscisic acid is crucial in all the stages of somatic development, maturation and hardening.

Synthetic Seeds

The concept of production and utilization of synthetic seeds (somatic embryo as substitutes for true seeds) was first suggested by Murashige in 1977. Synthetic seeds can be produced either as coated or non-coated, desiccated somatic embryos or as embryos encapsulated in hydrated gel (usually calcium alginate).

Successful utilization of synthetic seeds as propagules of choice requires an efficient and reproducible production system and a high percentage of post-planting conversion into vigourous plants. Artificial coats and gel capsules containing nutrients, pesticides and beneficial organisms have long been thought as substitutes for seed coat and endosperm. However, this technology is still in the developmental stage, and currently cannot compete with the other methods of commercial plant propagation.

Process of Micropropagation

The process of plant micro-propagation aims to produce clones (true copies of a plant in large numbers).

The process is usually divided into the following stages:

Stage 0   – Pre-propagation step or selection and pre-treatment
            of suitable plants.

Stage I – Initiation of explants-surface sterilization, establishment of mother explants.

Stage II – subculture for multiplication/proliferation of explants.

Stage III – shooting and rooting of the explants.

Stage IV – Weaning/hardening.

These stages are universally applicable in large-scale multiplication of plants. The individual plant species, varieties and clones require specific modification of the growth media, weaning and hardening conditions.

A rule of the thumb is to propagate plants under conditions as natural or similar to those in which the plants will be ultimately grown ex-vitro. For example, if a chrysanthemum variety is to be grown under long day-length for flower production, it is better to multiply the material under long-day length at stages III and IV. There is a wide option to undertake production of plant material up to a limited number of stages. For example, many commercial tissue culture companies undertake production up to Stage III, and leave the remaining stages to others.

Pre-Propagation Stage

The pre-propagation stage (also called stage 0) requires proper maintenance of the mother plants in the greenhouse under disease- and insect-free conditions with minimal dust. Clean enclosed areas, glasshouses, plastic tunnels, and net-covered tunnels, provide high quality explant source plants with minimal infection. Collection of plant material for clonal propagation should be done after appropriate pretreatment of the mother plants with fungicides and pesticides to minimize contamination in the in vitro cultures. This improves growth and multiplication rates of in vitro cultures.

The control of contamination begins with the pretreatment of the donor plants. They may be prescreened for diseases, isolated and treated to reduce contamination. The explants are then brought to the production facility, surface sterilized and introduced into culture. They may at this stage be treated with antibiotics and fungicides as well as anti-microbial formulations, such as PPM. The explants are then culture indexed for contamination by standard microbiological techniques, which are occasionally supplemented with tests based on molecular biology or other techniques.

Stage I: This stage refers to the inoculation of the explants on sterile medium to initiate aseptic culture. Initiation of explants is the very first step in micropropagation. A good clean explant, once established in an aseptic condition, can be multiplied several times; hence, explant initiation in an aseptic condition should be regarded as a critical step in micropropagation. More than often, explants fail to establish and grow, not due to the lack of a suitable medium but because of contamination. The explants are transferred to in vitro environment, free from microbial contaminants. The process requires excision of tiny plant pieces and their surface sterilization with chemicals such as sodium hypochlorite, ethyl alcohol and repeated washing with sterile distilled water before and after treatment with chemicals. After a short period of culture, usually 3 to 5 days, the contaminated explants are discarded.

The surviving explants showing growth are maintained and used for further subculture. In herbaceous plants *e.g.* potato, chrysanthemum, carnation, streptocarpus, strawberry, and African violet; the explant sources are meristems, apical- and axillary buds, young seedlings, developing young leaves and petioles, and unopened floral buds. The following low cost options can be adapted to initiate explants:

Sterile Instrument Technique: This method assumes that most of the deep-seated meristems and those covered by leaves or other integuments (*e.g.* floral bracts) are sterile. In this procedure, the explant is washed with sterile water, rinsed in ethanol, and instruments are sterilized every time they touch the surface of the explant, and the explant is moved to a new location on the dissection stage.

Surface Sterilization Technique: This is by far the most commonly used method. The explants are washed in sterile water, rinsed in ethanol, and surface sterilization is achieved by using chemicals with chlorine base. Calcium or sodium hypochlorite based solutions, 1–3 per cent (v/v) are usually used for soft herbaceous materials. A cheap and ready-made sterilant is 5-7 per cent solution of 'Domestos'- a toilet disinfectant which contains 10.5 per cent v/v sodium hypochlorite, 0.3 per cent sodium carbonate, 10.0 per cent sodium chloride and 0.5 per cent (w/v) sodium hydroxide and a patented thickener). The explants are washed in sterile distilled water before and after sterilization. Other surface sterilants used include mercuric chloride (avoid its use as far as possible, since it is highly toxic), hydrogen peroxide, and potassium permanganate.

For Soft Tissues:

- Wash explants from perennial plants for 1-2 hr in tap water. Eliminate this step for material from glasshouse grown plants.
- Wash in sterile distilled water three to four times for 5 to 10 minutes each.
- Dip in 95 per cent ethanol for 3 to 5 seconds.
- Wash once again with sterile distilled water for 5 minutes.
- Surface-sterilize in 5 per cent 'Domestos' (v/v) for 20-25 minutes.
- Wash with sterile distilled water three times for 10 minutes each.
- Drain water droplets by placing on pre-sterilized blotting paper.
- Transfer explants singly to the medium.

N.B.: Sterilize forceps each time to transfer explants to avoid cross-contamination.

For woody stems (e.g. roses, hardy shrubs, and trees):

- Collect stems, shoots, buds and store at 5 oC till needed.
- Rinse in ethanol for 3 to 5 seconds.
- Rinse in 1- per cent sodium hypochlorite (20 per cent bleach) for 10 minutes.
- Place lower parts of stems in flasks in 2 per cent sucrose and 200 PPM 8-hydroxyquinoline citrate at 23+2 oC. For items collected in September/October, add 50-PPM GA3.

After that 10 PPM GA will help break the dormancy:

- Re-cut the bottom of stem and replace the solution after 2 days.
- Excise the softwood from the developed shoots and use material for explants or for rooting.
- Surface-sterilize as in the above protocol.

Do not forget to sterilize forceps and scalpel every time for the transfer of explants to fresh solutions. Use sterile containers in the protocol of surface sterilization. If explants become brown or pale at the end of the protocol, reduce the strength of 'Domestos' to 2.5 per cent. Alternatively, dip explants in 10 per cent 'Domestos' for 2 minutes and then proceed to surface sterilize with 3-5 per cent 'Domestos' for 20 minutes. If basal contamination is observed after 2-3 days of culture, explants can sometimes be rescued by removing the basal end by making a single cut with a sharp scalpel and re-culturing on fresh medium.

Stage II: Stage II is the propagation phase in which the explants are cultured on the appropriate media for multiplication of shoots. The primary goal is to achieve propagation without losing the genetic stability. Repeated culture of axillary and adventitious shoots, cutting with nodes, somatic embryos and other organs from Stage I leads to multiplication of propagules in large numbers. The propagules produced at this stage can be further used for multiplication by their repeated culture. Sometimes it is necessary to subculture the in vitro derived shoots onto different media for elongation.

Stage III: The in vitro shoots obtained at Stage II are rooted to produce complete plants. If the proliferated material consists of bud-like structures (*e.g.* orchids) or clumps of shoots (banana, pineapple), they should be separated after rooting and not before. Many plants (*e.g.* banana, pineapple, roses, potato, chrysanthemum, strawberry, mint, several grasses and many more) can be rooted on half-strength-MS medium without any growth-regulators. Good sturdy well-rooted plants are essential for high survival during weaning and later transfer to soil. This stage is labour intensive and expensive. The process of in vitro rooting has been estimated to account for approximately 35–75 per cent of the total cost of production. Efforts should be made to combine rooting and acclimatization stages.

Stage IV: At this stage, the in vitro micropropagated plants are weaned and hardened. This is the final stage of the tissue culture operation after which the micropropagated plantlets are ready for transfer to the greenhouse. Steps are taken to grow individual plantlets capable of carrying out photosynthesis. The hardening of the tissue-cultured plantlets is done gradually from high to low humidity and from low light intensity to high intensity conditions. If grown on solid medium, most of the agar can be removed gently by rinsing with water.

Plants can be left in shade for 3 to 6 days where diffused natural light conditions them to the new environment. The plants are then transferred to an appropriate substrate (sand, peat, compost, etc.), and gradually hardened. Low-cost options include the use of plastic domes or tunnels, which reduces the natural light intensity and maintains high relative humidity during the hardening process. If the plants are still joined together after rooting, these should be planted as bunches in the soil and separated after 6 to 8 weeks of growth.

Delivery to the Growers

The delivery of the rooted and hardened small micropropagated plants to growers and market requires extra care. In some cases, plant losses can occur during shipment and handling by growers. This is particularly true when the plants are not fully hardened and rooted or not grown for sufficient duration after transfer to soil. Growers should be given clear instructions how to handle the material provided.

Apart from the economic loss, poor survival of planted material erodes the confidence of growers in the technology. The transfer of individual plants to soil in black plastic or polythene bags is widely used as a low-cost option to provide fully-grown banana plants directly to farmers in many developing countries.

Chapter 5

Genetic Resources and Plant Breeding

In order for genetic resources to be efficiently utilized in plant breeding programmes, it is first necessary to determine whether useful genetic variation exists in the material and secondly to develop the most cost-effective method of introducing the potentially useful genes into commercially acceptable material.

Until recently, genetic analysis has relied on conventional segregation analysis in controlled crosses for qualitative traits determined by major genes, and on standard biometrical methods for quantitative traits controlled by polygenes. Now, however, the use of DNA techniques has opened up new possibilities for genetic analysis and breeding, and in this chapter the general principles and applications of these molecular marker-based procedures will be discussed.

In its specific sense, marker-assisted or marker-aided selection (MAS) is the use of appropriate easily recognizable genetic markers to facilitate or accelerate the selection of linked genes controlling useful agronomic traits. The nature of the marker loci themselves is not important and they are used simply to indicate the presence of useful alleles of genes of commercial or practical importance. However, the term has also come to include the wider use of markers to improve breeding programmes.

Uses of Marker-aided Selection

In an ideal world, selection should be aimed directly at those genes which control the trait to be improved, but this assumes that genotypes at these target genes can be recognized easily, unambiguously and at an appropriate time without impeding the breeding programme. There are several reasons for using marker genes to improve selection efficiency.

Difficulties in Identifying Trait Genotype

The genotypes at a single-trait locus may be difficult to recognize because the desired alleles may have poor penetrance, be recessive, or interact with other genes or the environment. Environmental variation itself may impede accurate genotyping by causing the phenotype of different genotypes to overlap; this is particularly a problem with polygenic traits.

Some phenotypes such as resistance to a particular pest, pathogen or abiotic factor, like drought or salinity, may only appear under particular conditions which are difficult to define or control. These difficulties may be more severe during the early stages of a breeding programme when plant numbers are too few to allow adequate replication or the breeder does not want to place his valuable yet scarce breeding material at hazard.

Earlier Selection

Another major reason why it may be valuable to use marker or indicator genes is to reduce the time from sowing to selection. Many traits of economic importance are only apparent in the mature plant and so may not occur for months or even years after sowing. This is particularly a problem with biennial species and long-lived crops such as trees. Indicator genes, on the other hand, may be detectable in seedlings within days of sowing, if not in the seeds themselves, so avoiding the waste of valuable resources involved in raising and harvesting plants, most of which may prove to be of no value to the breeding programme.

More Intense Selection

Selection at a juvenile stage, particularly among seedlings, may also allow much larger populations to be studied and hence more intense selection to be applied. Indeed, selection may be possible in tissue culture without raising plants or running trials.

Nondestructive Scoring

Many characters are scored before maturity and the act of scoring may prevent seed being collected. Imposition of selection for pests and diseases may result in plants which are less able to produce seed. The use of certain types of marker loci may only require the removal of small quantities of leaf or other material to permit full genotyping of the plant.

Linkage Drag

Conventionally plant breeders have used backcrossing with selection to introduce a useful allele, such as one conferring disease resistance, from one strain into another. The source of the allele is often an alien or exotic species, but may be a related cultivar.

The F1 is backcrossed to the recurrent parent, and backcrossing is repeated for five to ten generations, the intention being to introduce the desired donor allele whilst returning to the recurrent parent's genotype at all other loci. At every backcross generation, the breeder selects for the phenotype of the allele to be introgressed whilst simultaneously selecting for the phenotype of the recurrent parent for all other traits.

It was shown by Stam and Zeven that this procedure results in the leaving of very large amounts of donor genome associated with the selected alien allele. On average, about 20% of an average chromosome will consist of a donor segment even after ten backcross generations, while in any given line the actual size could be much more.

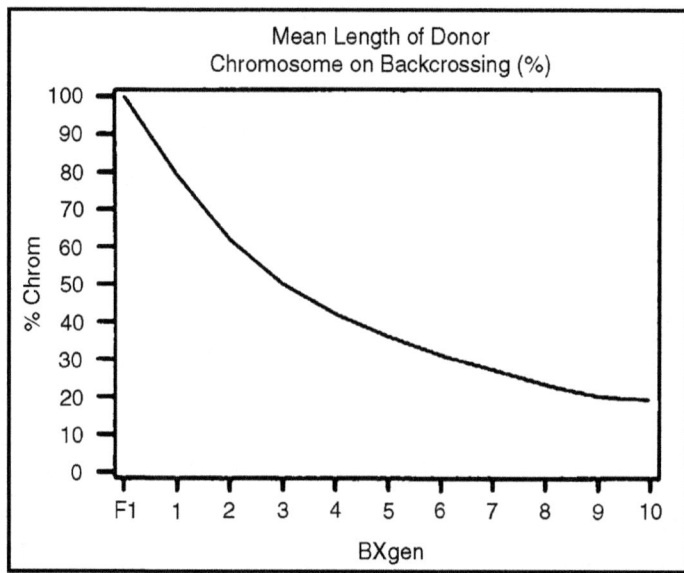

Figure 1: Linkage Drag.the Average Amount of Chromosome Surrounding a Selected GeneFollowing Backcrossing. Bxgen=Generation of Backcrossing from F_1

Thus, as well as introducing useful alleles, the breeder will also be introducing very many undesirable alleles from the donor and so the gain in one trait may be at the expense of losses elsewhere. The use of marker loci can enable the breeder to reduce the size of this unwanted associated region as well as to accelerate the speed of return to the desired genotype of the recurrent parent.

Heterosis Potential

Heterosis, or hybrid vigour, is due mainly to the presence of dispersed dominant alleles in the parental cultivars. Breeders concerned with developing lines for hybrid production are often concerned to identify lines which differ at as many gene loci as possible. It is often assumed that this implies that the lines also have very different genotypes at marker loci and hence such loci can be used to assess the genetic distance between potential parents, but this can be a dangerous assumption unless a wide range of different types of markers are used.

Quality Control

A major problem faced by most conservationists and breeders is to maintain the integrity of the material they are working with and to keep it free from contamination. Contamination can occur in various ways. It may arise through errors due to carelessness in labelling or sorting seed or plant material, mistakes in maintaining the pedigree, or as a result of outcrossing due to stray pollen from unrelated material.

Such errors are difficult to recognize when only the gross phenotype is available as a guide, but genetic markers can provide a clear genetic fingerprint which can be used at any time to confirm the origin of the material. It can also be used to confirm that the intended cross or self has occurred, or to separate the desired progeny from a particular parent which might be a mixture of selfs and crosses.

Other Uses

Marker information can be used in other situations also: distinguishing haploids, diploids and aneuploids, for example, particularly following wide interspecific crosses or tissue culture. Doubled haploid lines produced by microspore or ovule culture can be recognized in this way and possible aneuploids arising during chromosome doubling can be identified and removed. Not only can genetic markers identify contaminant material but they also allow closely related or identical material to be identified in genebank collections, so reducing unnecessary duplication.

Types of Genetic Markers

Up to the mid-1960s the only easily usable genetic markers were those that produced clearly visible effects on the plant's phenotype such as colour, size, shape, disease resistance, etc.. All these were the result of mutations to alleles which had, obviously, a major effect on the development of the plant. Such mutants were rare and generally not desirable in commercial material although they may be selected by breeders of decorative plants for their novelty value.

As genetic markers, however, they have many disadvantages. They often affect the fitness of the individual and could well have effects on important agronomic traits, *i.e.* they exhibit pleiotropy. Because they are rare, it would be necessary to bring several such mutants together in the same material in order to use them, and they would not normally be present in commercial cultivars or, indeed, in wild material. This not only takes time, but the combined effects of many such alleles would mitigate against their value as a genetic tool.

In the 1960s through to the 1980s, naturally occurring genetic polymorphisms at the protein level came to be easily recognized. Variation in enzymes and storage proteins was detectable on an electrophoretic gel for a very high proportion of gene loci which could be studied in this way and, unlike major mutants, such naturally occurring allelic variants had much smaller effects on fitness and hence were more common and existed naturally in populations. The number of such polymorphisms was, however, insufficient to provide the coverage of the genome desired by breeders and conservationists. The development of techniques during the late 1980s up to the present revealed the vast amounts of genetic polymorphism which exists at the DNA level and has revolutionized genetic analysis, opening up a whole new range of genetic tools.

As described elsewhere in this book, this variation arises through the existence of occasional base changes in the DNA which can be recognized by restriction enzymes or by primers used in analysis involving the polymease chain reaction (PCR). Much of the DNA, possibly as much as 90% in many species, is noncoding and hence reliance on morphological or biochemical variants was only using the restricted variation that plants could tolerate within the coding parts of their DNA.

Clearly, there are limits to the tolerance of such variation. Variation in the noncoding regions, whether in the intergenic regions or within introns, is probably under far less constraint by natural selection, and hence the number of polymorphic sites is potentially enormous. Because most of these sites are outside coding genes they are referred to as loci rather than genes; genes are coding regions whilst most marker loci are not. Depending on how the DNA polymorphism is studied, various types of marker are used. Restriction fragment length polymorphisms (RFLPs) have been the most widely used until now. They have the advantage of being codominant and hence all genotypes in a cross can be identified. On the other hand they require quite large quantities of DNA, and hence plant material, and generally rely on radioactive probes, although fluorescent techniques are becoming more popular.

As a result, RFLPs are expensive to use. The use of radioactive probes incurs safety considerations, as well as taking longer to visualize the bands. PCRbased markers are now being used more widely.

Randomly amplified polymorphic DNAs (RAPDs), for example, require far less material because only small amounts of DNA are needed and the required sequences are then amplified. They do not involve radioactive probes and are both quick and cheap. On the other hand RAPDs are dominant although it is possible to distinguish heterozygotes from homozygotes by the amount of product. Amplified fragment length polymorphisms (AFLPs) offer even cheaper and more easily identified and extensive PCRbased polymorphism. Because of the number of polymorphisms which exist on individual gels and the speed of production, computer software is necessary to scan the banding patterns on the gels and to analyse and assimilate the information that is generated. Other popular markers are microsatellites (short sequence repeats) and cleaved amplified polymorphisms (CAP)-based primers. All the genetic variants discussed above will be subsumed under the broad title of marker loci. They may be of little interest in their own right, and this is particularly true of the molecular polymorphisms; their value lies in their use as markers of more useful genes.

Gene Mapping

An important, though by no means an essential, step in genetic analysis is to produce genetic maps of the marker loci. Such maps are often referred to as 'framework maps' because they provide a framework within which important genes can be located, as well as providing a means of comparing chromosome organization in other closely or distantly related species. There are two stages to mapping. Firstly, to arrange the markers in a linear sequence separated by an appropriate map distance, *i.e.* to construct a linkage map. The second is to relate the linkage maps to particular recognizable chromosomes. The latter is often the most difficult and generally not essential for breeding or conservation work. We will, therefore, concentrate on the former. Chromosomes contain a single linear molecule of DNA and hence the markers on that chromosome occur at particular positions along that molecule. Typically, each chromosome contains 107 to 108 base pairs (bp), *i.e.* 104 to 105 kilo-base pairs (kbp) of DNA, while a typical structural gene, coding for a polypeptide chain, would be between 1 and 2 kbp long. Assuming only 10% of the genome is coding DNA, then a chromosome probably contains something of the order of 1000 to 10 000 genes. Fortunately, the fact that the chromosome is a linear molecule means that the genes need to be mapped in only one dimension.

Currently, the only useful method of gene mapping, at least as far as breeders and conservationists are concerned, relies on recombination between homologous chromosomes resulting from genetic exchange, which can be seen as chiasmata at diplotene through first metaphase of meiosis. A single chiasma on a particular chromosome results in half the gametes from such a meiosis being recombinant for that chromosome: half the gametes will contain a copy of that chromosome that is part maternal and part paternal in origin, while the remaining gametes will be entirely parental. Very few plant species have a sufficiently low number of clearly recognizable chromosomes that the number of chiasmata on a particular chromosome can be compared in different nuclei.

However, in many species the total number of chiasmata in each diplotene nucleus can be counted, at least in pollen meiosis, and the average number of chiasmata per nucleus calculated. The number and position of chiasmata on a particular pair of homologous chromosomes will vary from nucleus to nucleus; it is as though there are a large number of potential sites of exchange, but that in any given meiosis only a few sites are actually involved.

The longer the chromosome the more potential sites there might be. There will invariably be one chiasma and there may be two, three, four or more; a typical set of results is shown in Table A chromosome that has one chiasma on average is said to be 50 centimorgans (cM) long – a map length unit named after the American geneticist Thomas Hunt Morgan.

This relates to the fact mentioned above that one chiasma results in 50% recombinant chromosomes. By extension, a chromosome with an average of 2.5 chiasmata is said to be 125 cM (*i.e.* 2.5 350 cM) long. It appears that, as a rough rule of thumb, each chromosome has on average two chiasmata and so is 100 cM long. It follows, therefore, that providing the haploid chromosome number is known (*n*), the total map length will be approximately 2 3 *n* 3 50 cM. The average number of chiasmata per nucleus has been calculated in many species and this rule generally holds.

Table 1: Chiasma Fraquency and Mapping.Chiasma Frequency in Chromosomes of Secale Cereale.

Number of chiasmata per chromosome	Frequency	Percentage unrecombined Chromosomes
1	0.267	50.0
2	0.716	25.0
≥3	0.017	12.5
Mean	–	22.0

This rough rule is important because it gives the geneticist an idea of the total genetic map length that should be expected as more genes or markers are mapped. It provides a guide to the extent to which the currently mapped markers cover the full genetic map. Because the map is based in chiasma units, it is not the same as a physical map. The distribution of chiasmata is not uniform over a given chromosome in a species and will vary with the chromosome and the species.

Although a knowledge of the number of chromosomes and, ideally, the mean chiasma frequency per nucleus for the species, indicates the total map length, the actual map has to be constructed from examining the frequencies of progeny in crosses; *i.e.* maps are constructed from information based on the consequences of chiasmata, namely recombination of genetic markers. Recombination of genetic markers is identified by the presence of gametes that contain recombined markers, and such gametes can be recognized from the phenotypes of progeny of a cross.

The frequency of such gametes is an estimate of the recombination frequency between the two markers concerned. The closer two markers are to each other on the chromosome, the less likely a chiasma will occur between them and the less likely they are to recombine. Providing the two markers are sufficiently close that either none or one chiasma occurs between them but never more, then the recombination frequency as a percentage is equal to the map distance in centimorgans as described above.

The problem arises when the markers are sufficiently far apart for two or more chiasmata to occur between them in some meioses. It can be shown that not only does a single chiasma between a pair of markers result in 50% recombination between them, but so also do two, three or more chiasmata, on average. It is necessary to say 'on average', because there are a variety of possible consequences with two, three or more chiasmata in any given meiosis but, in practice, when one is looking at the progeny of a cross or self, each progeny will be the result of gametes from different meioses.

This fact implies that although map distance increases linearly with the number of chiasmata, the same is not true with the frequency of recombination, which can never be more than 50% for a pair of markers even though they might be 100 cM apart at opposite ends of a chromosome.

These relationships are illustrated in Figure. In order to overcome this problem, mapping functions have been devised to correct for multiple chiasmata. The most common of these are the Haldane (1919) and Kosambi (1943) mapping functions. Haldane's assumes that the probability of none, one, two or more chiasmata in a given interval follows a Poisson distribution, i.e. that the chiasmata are independent, random events. Kosambi's method allows for a certain degree of nonindependence in chiasma occurrences. It has long been known by cytogeneticists that chiasma interference occurs over quite large regions of a chromosome.

Interference means that if a chiasma occurs at a particular point on the chromosome, the next one will never occur closer than some fixed distance, the interference distance, from it. Beyond this distance, there is a short distance in which the interference disappears and the next chiasma can then occur randomly outside this range. Observational data in various species of plants are confirming this and suggest that the interference distance is generally between 15 and 20 cM, i.e. 15 to 20% of a typical chromosome on either side of the chiasma. This has important and useful consequences for genetic and breeding work as will be shown later.

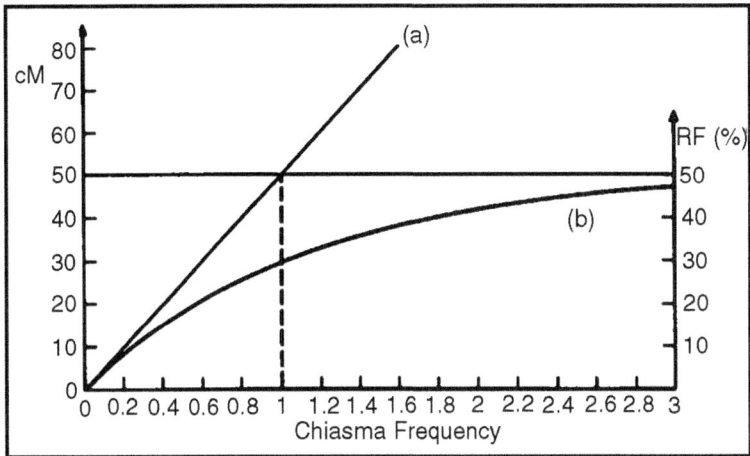

Figure 2: The Relationship of Chiasma Frequency With (a) Genetic Map Length and(b)Recombination Frequency.

Recombination frequencies between pairs of markers can be scored in a wide variety of different crosses. The simplest to use are generations derived from an F1 because only two alleles are segregating, and their distribution in the chromosomes of the parents of the F1 can be determined. The generations that can be used are F2, backcrosses (Bc), recombinant inbred lines (RILs) or doubled haploid (DH) lines, with F2s being the most informative.

General formulae for calculating recombination frequencies are given by Allard. In outbreeding species where F1s are not available, a given individual can be considered as an F1 and its selfed progeny as being an F2, even though the inbred parents do not exist. If it cannot be selfed but controlled crossing to another individual is possible, then again those genes segregating in the cross can be mapped although the situation is more complex.

There could be from two to four alleles segregating at each locus and, with respect to any one locus, the cross could represent an F2, Bc1 or Bc2. Moreover, the distribution of alleles in the two parents cannot be known and has to be inferred from the progeny. This complexity is compounded by the large number of marker loci that may be segregating in any given cross. Fortunately, a range of software packages are available to estimate these recombination frequencies, identify linkage groups, assign the markers to the most likely order and space them in map units (cM) on these linkage groups; examples are MAPMAKER and JOINMAP. Ideally the number of linkage groups should be equal to the chromosome number in the gametes but, unless the markers available provide a good coverage of the genome, it is likely that those markers on a given chromosome may appear as two or more separate linkage groups simply because the subsets are not sufficiently close for them to be recognized as being together on one chromosome.

A typical marker framework map is shown in Figure for the chromosomes of *Brassica oleracea*. Clearly these maps provide a very detailed coverage of the chromosome. Many of the markers are very close, *i.e.* less than 10 cM, and over this distance it matters little which of the two mapping functions is used. With the more widely spaced markers, *i.e.* recombination frequency >15%, Haldane's function will exaggerate their distance apart because it will allow for more double crossovers than actually occur.

As more markers are mapped, the total length should converge on the value predicted by the chiasma frequency, *i.e.* approximately 2 3 n 3 50 cM. It is important to remember that genetic map distances have standard errors which are dependent on the size and type of population used to construct the map, the mapping population. The map obtained is that which best fits, given the data and the assumptions underlying it, but there may well be other maps which also fit the data well, though are slightly less likely. Any particular map always develops a greater aura of respectability once it is published, frequently without stating its reliability, and subsequent yet different maps are treated with suspicion. A recombination frequency of true value p has a standard error of $?[p(1\ 2\ p)/N]$, where N is the size of the gamete population. For example, if p is 0.1 (*i.e.* RF = 10%) and N = 100, the standard error is 0.03 or 3%. In other words the 95% confidence interval of the estimate lies approximately between 4% and 16%.

There are other reasons, apart from statistical sampling, why maps could be wrong. The most obvious is the accuracy with which the data are scored and recorded. Autoradiographs and banding patterns on gels can easily be misread and research workers are loath to discard data even though its interpretation is ambiguous. All data should be checked independently by two or more people and any ambiguities removed.

Wrong scores will bias the data and often exaggerate map lengths as they suggest double recombination; wherever double recombination appears to have occurred in two short, adjacent intervals, the data should be checked. Changes in methylation rather than scoring errors could cause a restriction site to appear or disappear so giving the false impression of double crossing- over, or it could simply be an error.

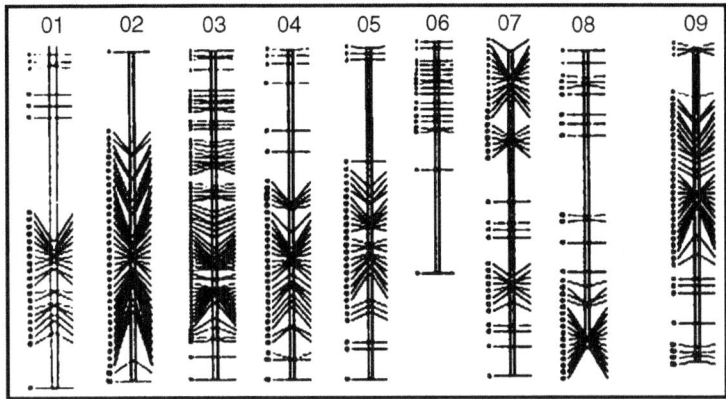

Figure 3: Typical Molecular Marker Framework Map. The Nine Groups of Brassica Oleracea Based on RFLP, Isozyme and Morphological Markers

Where maps appear to have major inconsistencies in different crosses, different chromosomal structural arrangements may be responsible, such as translocations or inversions. There is considerable evidence that chiasma frequencies, and hence recombination frequencies, may be quite different in male and female meioses even on the same plant although in other studies this may not be so.

Thus although the order of markers on a linkage group may remain the same, the relative distances between them may be quite different if the map derives from male versus female meioses. For example, in *Brassica* it appears that the genetic map obtained from a backcross where the F1 is the female parent is 60% longer than when the F1 is the male parent. It is also clear that environmental factors during meiosis, particularly temperature, can affect the number and distribution of chiasmata, and this could be very critical if the crosses are set up at different times of the year.

Two or more different populations will almost certainly be segregating for different combinations of polymorphic markers, although some at least should be in common. These common ones can be used to overlay the two or more maps and a consensus map produced from an amalgamation of these. Again software is available to produce these consensus maps. Despite the various error-causing factors discussed above a considerable degree of consensus can be produced from such populations and, considerable similarity of chromosome sequence is conserved between even distantly related species, allowing the potential for cross-species genetic transfer.

It was stated above that the total map length should be dictated by the chiasma frequency, in so far as this can be accurately measured. When the first few genes were initially put onto genetic maps in the first 40 years of this century, they only covered a small part of the genome – except in wellstudied species such as *Drosophila* and maize. As more and more genes were placed on the map, so the total map lengths increased towards the asymptote dictated by the chiasma frequency. However, during the first decade following the use of molecular markers, the map lengths of some species appeared to exceed that expected, and it was even argued by some that the chiasma theory of recombination might be wrong.

No one would wish to argue that chiasma frequency data are free from error but there would have had to have been quite excessive underscoring to result in the disparity which was occasionally found. It would appear, however, that the excessive lengths were due in part to errors in scoring and to the use of small mapping populations, because subsequent map sizes have shown a progressive decrease towards the predicted asymptote. Scoring errors bias estimates of RF generally upwards, while RFs have to be large to be detectable in small populations. Our own experience with mapping in cereals and brassicas has also shown a progressive decrease in estimated map length as more markers can be found to fill some of the large gaps in the map and as errors are removed from the mapping data.

Locating Genes of Major Effect

The use of extensive molecular and other markers as described above provides a general framework map. Although this has value in its own right for comparing linkage groups in different species as well as possible structural variation within a species, the main value lies in providing a set of markers to locate genes of economic or special scientific interest. Locating individual major genes is a straightforward development of the procedures above. Let us assume that a cultivar is identified which contains an allele of interest which changes the phenotype in a clearly recognizable fashion and which appears to segregate as a single gene in crosses to other cultivars. Using a knowledge of the positions of existing marker loci, a small subset of markers can be chosen which provide a reasonably even coverage of the genome, say every 20 to 30 cM, *i.e.* approximately five per chromosome. These then have to be shown to differ between the two cultivars and any which are monomorphic replaced by polymorphic markers.

A rough position of the gene of interest can then be obtained by bulked segregant analysis. This involves taking the mapping population, *e.g.* a Bc, and dividing the plants into two groups depending on whether or not they show the phenotype of interest. Bulk samples of DNA are taken from each group, and the two samples assayed for the marker systems chosen as in the paragraph above. Any marker which is unlinked to the gene to be located will produce similar banding patterns in both samples because it will be segregating independently of the target gene. Conversely, should one of the markers be very closely linked to the target gene, it will co-segregate and the two samples will show quite different patterns for that marker.

In practice, complete co-segregation with any one marker is not very likely but, given a reasonable spread of markers over the genome, one or two are likely to show a strong association with the two groups. Thus although bands associated with both marker alleles will be found in each sample, their relative intensities will be quite different because of the association with the target gene. Having located the chromosome and the region on the chromosome, the actual position can be identified in a segregating population using the principal marker associated in the bulked segregant analysis together with other polymorphic markers situated about 10 cM on either side of the principal marker.

The target gene is then mapped with respect to the three markers. This approach can be used for a wide variety of populations and traits. Even if the trait has poor penetrance, the affected group will clearly show the marker banding pattern even if the nonaffected group consists of a mixture due to misclassification.

Mapping Genes Controlling Quantitative Traits

Quantitative traits such as yield, quality, height and flowering time, which are controlled by several genes and are greatly influenced by the environment, create particular difficulties for gene mapping. The problems arise because the genotype for the trait concerned can never be clearly identified from the phenotype; many different genotypes could produce the same or very similar phenotypes whilst the same genotype can result in different phenotypes depending on what may be elusive factors in the environment.

Whereas with a single gene difference controlling a major effect, two or more Mendelian phenotypic classes might be recognized in ratios of 3:1, etc., quantitative traits resulting from the joint segregation of many genes show a continuous, often normally distributed range of phenotypes. The total phenotypic variation in an F2, for example, VP, is made up of genetic and environmental components, VG and VE. The genes underlying such traits are variously called polygenes, effective factors or, more recently, quantitative trait loci or QTL.

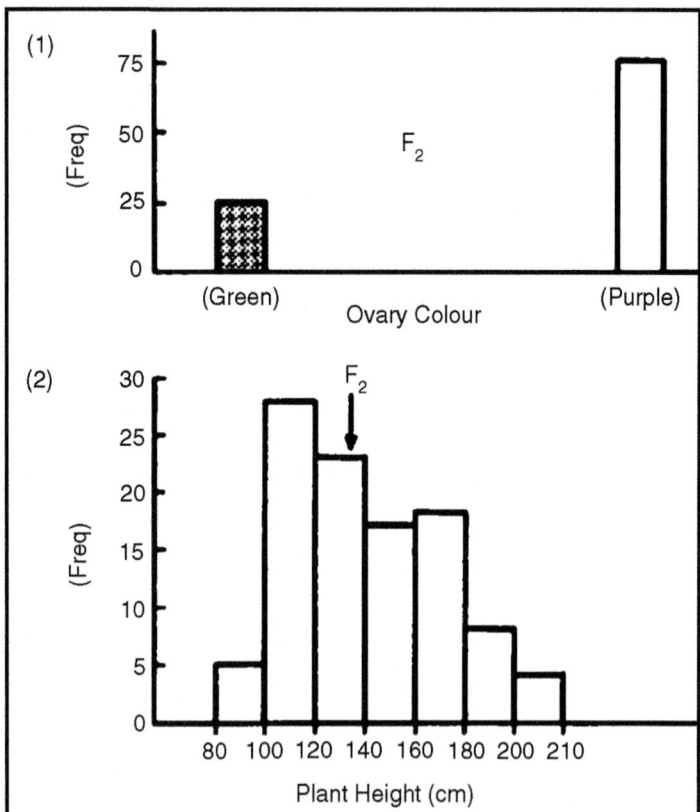

Figure 4: Contrast Between Distribution of a Qualitive and Quantitative trait in an F_2. Data for NIcotiana Rustica,Ovary Colour and Height.

However, it is still possible to map and measure the effects of the QTL by techniques analogous to those used for single major genes, *i.e.* by looking for the effects of co-segregation of particular marker loci with differences in the quantitative trait. Various mapping populations can be used as before, F_2, Bc, RIL, DH or open-pollinated full sib (FS) or half sib (HS) families, and the individuals or families are scored both for their phenotype for the quantitative trait(s) and for their genotype at the marker loci. The type of population that is used will depend on the traits studied and the breeding system of the plant species but, other things being equal, an F_2 population will normally provide the most information for a given size.

Unfortunately, most traits of economic importance such as yield, quality and disease resistance are not usefully scored in an F_2 population because one is interested in how the material performs in something similar to the high-density monoculture that it will encounter in agricultural practice. Measuring yield in an F_2 population with spaced plants is straightforward and statistically very powerful but it may well not provide any insight into how yield is controlled in agricultural practice.

It is therefore necessary to use a plot structure for most traits in an attempt to simulate commercial conditions. For this reason most plant breeders will prefer to work with genotypes that can be extensively and easily replicated such as RILs, DHs or F_3s. It is also easier to explain the principles underlying QTL mapping using a population of DH lines, and so for both reasons the methodology will be illustrated with DH lines, although the same principles apply to all populations. Because several trait loci are concerned, probably on different chromosomes, bulk segregant analysis is of limited applicability, at least initially.

Theory of QTL Mapping in Doubled Haploid Lines

Doubled haploid (DH) lines can be produced parthenogenetically from an F1 by microspore or ovule culture and, in some species, such as wheat and barley, by wide outcrossing. Each original DH plant is derived from a single gamete, which subsequently becomes homozygous and disomic either naturally or by treatment with the drug colchicine, which inhibits spindle formation, and hence chromosome disjunction, at metaphase of mitosis.

Selfed seeds from such a plant produce a DH line of identical individuals so the line becomes effectively immortal. If we consider an F1 that is heterozygous for a marker locus M1/M2 and for a linked QTL locus Q1/Q2, where the 1/2 indicates the allele with an increasing or decreasing effect on the trait, this will produce the gametes and hence the DH lines shown in Table. If the DH lines are scored for some trait for which the Q+/Q+ homozygotes increase the trait above the overall mean, m, by a units while the Q2/Q2 homozygotes decrease the mean by a, then the means of the four gametic types are as shown in Table, where R is the recombination frequency between the QTL and the marker. From this it follows that the mean trait score of all those DH lines which are M1/M1 (M1M1) is m 1 a(1 2 2R), while the mean of all those DH lines which are M2/M2 (M2M2) is m 2 a(1 2 2R). In other words, half the difference between these two means (M1M1 2 M2M2) is a(1 2 2R).

Clearly, if the marker is so close to the QTL that it never recombines, R will be 0 and the marker difference will represent the QTL effect, a. Conversely if the two loci are unlinked, *i.e.* R = 0.5, then the difference will be zero. This effect is, of course, the basis of bulked segregant analysis. In general, the magnitude of the marker effect will decline linearly with the distance of the marker from the QTL in terms of recombination frequency, and so with several linked markers, their individual trait effects will be as shown in Figure. Similar Arguments Apply to Other Populations Such as F_2s and RILs.

If the trait means of the markers and their map are known very accurately, and only one QTL existed on the chromosome, then as Figure shows, it would be relatively easy to locate the QTL. Figure shows the distribution of flowering time among a number of doubled haploid lines of *Brassica oleracea* containing alternative alleles at a marker located very close to a QTL. In practice neither the map positions of the markers nor their means are known with very great accuracy because the number of DH lines and replicated plots are generally few and the heritability of quantitative traits is generally low. Moreover, linked QTLs create difficulties because their individual effects can combine either to create the impression of an intermediate, ghost QTL, or to conceal their effects altogether.

Various analytical procedures have been developed to maximize the efficiency and accuracy of locating individual QTLs and to separate linked QTLs. They all rely on the relationships described in the previous paragraph and use either maximum likelihood or weighted least squares regression procedures to estimate the QTL locations and effects that best fit the observed trait scores for each genotype.

The main difficulty is that there are two unknown parameters with respect to each marker score; the QTL effect, a, and map position. This means that it is necessary, in all methods, to use an iterative approach which involves trying all possible QTL locations along each chromosome and identifying the position or positions of the QTL which best fit the observed data as indicated by the size of the likelihood or the residual regression variance. As a result, a very large number of statistical tests are performed with the concomitant risk of false positive results.

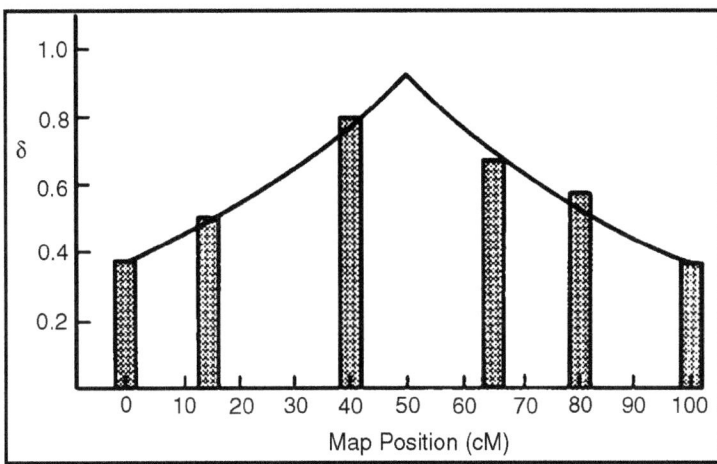

Figure 5: Quantitative Trait Effects (δ) Associated with Marker Loci.The Bars Indicate the Observed Effects,While the Curve Illustrates the Expected Relationship if a QTL of δ = 1.0 Existed at 50 cM

All methods provide estimates of QTL locations and effects with similar precision in terms of confidence intervals, and this precision drops rapidly as the heritability of individual QTLs declines, *i.e.* the estimates are accurate only when there are just two or three unlinked QTLs with large effects. For example, Hyne *et al.* showed that the 95% confidence interval for the location of a QTL contributing 10% to the phenotypic variance of an F2 could be as large as 35 cM.

However, conservationists and breeders who are concerned with introducing QTL into commercial cultivars from other cultivars or more distantly related species, would normally only be interested in those cases where a few QTLs of major effect are to be manipulated. There is no evidence that dense maps are required for initial

QTL location. It is better to have large populations and a few (5–10) markers per chromosome. There is evidence that many of the larger QTLs located so far map closely to previously known major genes. This is often used as evidence that there are, in fact, very few QTLs, but this may well be misleading for several reasons. The QTL map locations are sufficiently imprecise that they could well appear spuriously close to candidate loci.

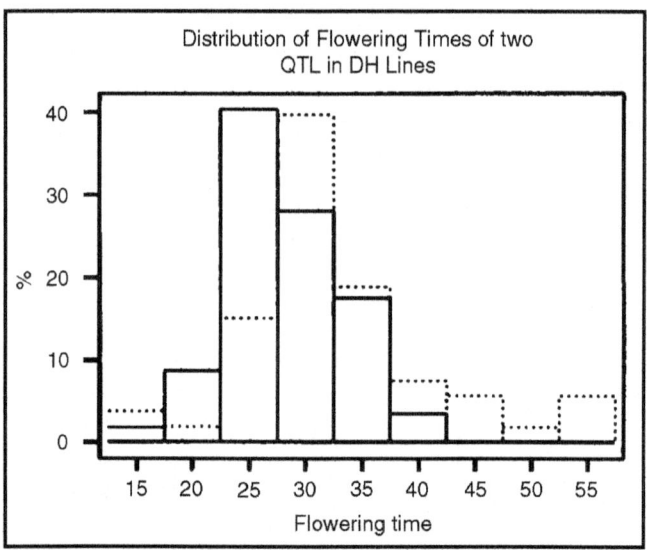

Figure 6: Distribution of Flowering Times Among Doubled Haploid (DH)Lines in Brassica Oleracea Containing alrernative Alleles of a Marker Linked to a QLT.The two sets Significantly Different Means Indicating a QLT with an Effect of Apporoximately ± 2 Days

Only QTLs of large effect will stand much chance of being located and, by definition, they are likely to be major genes. It is likely that many cases where QTLs have very large effects could be due to the chance association of alleles of like effect at several QTLs along a chromosome, which cannot be separated as individual effects. Much effort has been devoted to increasing the precision of QTL location, in particular in improving the power of the test to reduce the failure to detect genuine QTL.

A long-established problem with breeding for quantitative traits in plants is the genotype 3 environment interaction. Using QTL location techniques, it is now possible to explore these effects at the level of individual QTLs although it is often difficult to know whether genotype 3 environment or poor repeatability is responsible.

Alien Gene Transfer

Once useful genes have been located on a framework map, it is then possible to use the molecular markers to facilitate the transfer of the useful genes between species and strains in an efficient manner.

As stated earlier, recombination is not a frequent event along a chromosome and, furthermore, two recombination events rarely occur within a distance of 15 to 20 cM of each other.

It therefore follows that if a useful target gene (a major gene or a QTL) is known to be located within a region of chromosome flanked by two markers that are no more than 15 cM apart, then these markers can be used to follow and control the progression of the target gene through successive stages of a breeding programme.

Clearly, the alleles at the marker loci have to differ in the donor and recipient cultivars, but any chromosome that contains A1 and B1 at the marker loci will also contain T1, the target allele to be transferred. Should single recombination occur close to T, then the chromosome will contain A1 or B1 but not both. Providing the markers are less than 15 cM apart double recombination resulting in A1T2B1, and hence loss of T1, can be safely ignored.

If the location of T is not accurately known, then three or more marker loci may be needed to be sure of safely bracketing the region containing T without fear of double recombination. There are conflicting requirements in marker-assisted gene transfer: the need to keep the bracketed region large enough to be sure of holding the gene to be transferred whilst not having it so large that too many other linked but undesirable alleles are transferred with it. Let us now consider the basic procedure of gene transfer. Traditionally, breeders have introduced a useful allele into their cultivars by backcrossing the original F1 to the commercial cultivar whilst selecting at each generation for the desired allele.

As was shown earlier, this can result in a very large region of chromosome around the target gene surviving even after ten generations of backcrossing – a phenomenon known as linkage drag. With markers, however, it is possible to reduce linkage drag considerably whilst simultaneously reducing the number of backcrossing generations to two or three instead of the normal six to eight.

Moreover, the amount of plant material raised at each generation can be reduced because much of the molecular genotyping can be achieved at the juvenile stage. The aim is to hold the target allele, T1, heterozygous throughout the backcrossing process, by selecting for the flanking markers, whilst simultaneously selecting for the genotype of the recurrent, commercial cultivar at all other loci. The latter can readily be achieved by having as few as four or five well-spaced markers on all chromosomes and selecting for recurrent parent alleles at each generation.

Depending on the chromosome number, some individuals in the first backcross generation will be homozygous for several whole chromosomes whilst still being heterozygous for the target gene. Those which have most of the recurrent parent chromosomes are selected and backcrossed again, and if necessary the process is repeated for a third generation. For each generation, fewer markers need to be screened as those found to be homozygous in the previous backcross no longer have to be checked.

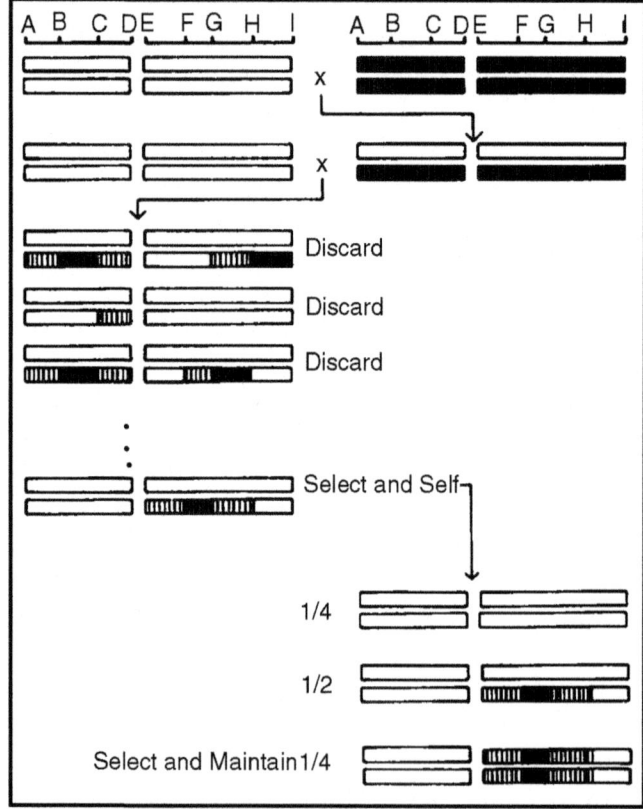

Figure 7: The use of Molecular Markers to Facilitate Gene Introgression.

Unless the position of the target gene is very accurately defined, it is prudent to have several markers covering its likely position, again with the proviso that they need to be separated by no more than 15 cM. At the final stage of the introgression process, a backcross individual which is homozygous for most of the recurrent parent alleles is selfed or, if this is not possible, intercrossed with a similar genotype.

From among the progeny, individuals are chosen which have different combinations of markers bracketing the target region as shown in figure. These can be selfed again and homozygotes for the different sequences identified and multiplied for reassessment for the target trait. Some will fail to show the target trait because of recombination, but of those that do, the lines with the shortest sequence of donor markers are selected for further trials. The approach can obviously be adapted to the simultaneous introgression of several genes: no new principles are involved but the screening process becomes more elaborate.

It is always possible at a later stage to reduce the length of the introgressed region around T by looking at other, more closely linked, markers. Similarly, it is wise to check the origins of the other chromosomes to avoid having introgressed unwanted regions through missing double recombinants in the central regions of individual chromosomes or single recombinants towards their ends.

If such errors are found they can easily be corrected by another round of backcrossing to the recurrent parent. When attempting to reduce the length of donor chromosome surrounding the target gene, it is important to realise that this should be done in two successive generations. If the sites of the required recombination events are less than 15 cM apart, then the simultaneous double event will not occur.

Even if they are far enough apart for the double event to occur, the probability of such an event may be too low to be practicable. If two recombination events each have a probability of 0.05 (i.e. 5%), their combined probability is at best 0.0025.

Therefore, it would be necessary to genotype approximately 1200 individuals to be 95% sure of having at least one double recombinant. If it was tackled in two stages then it would require only the genotyping of 86 (= 28 in the first round plus 58 in the second; less being needed in the first round because the recombination could occur on either side of T), a very considerable saving in effort of 93%, or 1114 plants! Marker-aided selection has been successfully tried for quantitative traits in several crops while the efficiency of the procedure has been examined by Lande and Thompson (1990).

Map-based Gene Cloning

Providing a target gene can be located to within a pair of marker loci, it should be possible to use standard cloning techniques to walk along the chromosome between the markers and to find the target gene. The feasibility of this depends on the distance between the markers and the ease with which the target gene can be recognized from among the clones.

Depending on the species, a genetic length of 1 cM can, on average, consist of as few as 280 kbp of DNA in rice, or 2220 kbp in maize. It is not always easy to locate a target gene with sufficient accuracy to be able to say that it is between two markers as close as 1 cM, and hence the actual lengths of DNA between them can often be much greater. With QTLs one can seldom achieve anything approaching this accuracy by existing methods of conventional mapping. Furthermore, these are average distances per centimorgan; some chromosomal regions could be much longer or, indeed, less.

Increases in mapping reliability can be obtained by using what are called near isogenic lines (NILs), which are lines obtained by selfing or backcrossing and are known to differ from some standard genotype by just a short defined section of chromosome. Providing this section is delineated by accurately mapped flanking markers, and the NILs which do and do not contain this region can clearly be shown to differ for the target gene, then the target gene has to be in that region. The genetic and physical proximity of the markers can be established by conventional mapping and by gene cloning.

Map-based gene cloning involves starting from one of the markers and walking to the next by means of a series of partially overlapping cosmid clones. Cosmid clones are necessary because they can be up to 50 kb in length and hence require fewer cloning steps in order to cover the DNA between the markers (280 to 2200 kb per cM in the examples cited above). Identifying the target gene is more difficult. Clearly, if the gene is between the two flanking markers it has to be on one of the cloned regions and could possibly be on two overlapping clones.

It could be identified by transforming plants with each clone separately and identifying which transformant expresses the effect. If the gene concerned exhibits classical dominance it would be necessary to transform the homozygous recessive with the dominant allele or the transformant would not be recognized. Alternatively, the gene has to be introduced into a species that does not normally express the gene. Once the effective cloned fragment is identified, the actual location of the gene within the fragment could be identified by successively transforming sub-fractions of the clone or, following sequencing, by looking for potential open reading frames. In those cases where the target gene produces a known product, cDNA clones derived from tissue likely to be rich in the appropriate mRNA of the target gene can be used to identify likely sequences in the previous overlapping clones.

Once the gene has been identified by one of these methods, its structure and activity can be studied in detail. Disease resistance genes are obvious candidates for such studies as their action is very specific and their location on the genome of many species is well documented. However, unless suitable candidate loci.

Cloning of Plant Traits

Plant gene cloning is currently limited to simply inherited single gene traits. The transfer of genes to plants is limited to single genes or, at most, two to three genes encoding a particular, well-characterized biochemical pathway. Manipulation of multigenic traits is currently only feasible within the existing gene pool, sometimes made more efficient by markerassisted selection.

When traits are regarded in the molecular genetic sense, often the coding sequence is the major consideration. Obviously, the coding sequence is of major importance as this is the template for determining the amino acid sequence of the final protein or enzyme which determines the phenotype. However, a major factor in controlling the phenotype is the presence of regulatory sequences surrounding the coding sequence in the form of upstream (promoters and enhancers) and downstream (terminator) regulatory sequences and, in many eukaryotic genes, introns.

When traits are being manipulated via genetic engineering, the role of these regulatory sequences cannot be trivialized. Put simply, these sequences control such things as the level of gene expression, tissue/organ specificity, developmental expression and response to particular abiotic/biotic stimuli.

The regulatory sequences that have been subjected to greatest attention in plants are the upstream promoters. For proper expression of a seed storage protein, for example, the gene of interest must be under the control of a promoter which expresses in developing endosperm (monocots) or cotyledons (dicots). Expression of other genes may be required only in anthers, root hairs or photosynthetic tissues, or in response to a wound such as that caused by a chewing insect or penetration of a fungal infection structure. In the earliest cases of transgenic plants, introduced genes were placed under the control of constitutive promoters of bacterial (octopine or nopaline synthetase) or viral (cauliflower mosaic virus (CaMV) 35S) origin. These promoters have been found to give adequate expression of selectable marker and reporter genes in dicotyledonous species.

However, expression is commonly not of a sufficient level in monocots, especially cereals, for good selection with antibiotic or herbicide resistance genes. And, as already discussed, such a constitutive pattern of gene expression is commonly not desirable for most traits. For cereals, a number of other promoters have been developed to achieve high-level constitutive expression of selectable markers or reporter genes. These include sequences cloned directly from cereal tissues such as the ubiquitin and rice actin 1 promoters and synthetic promoters involving various spliced sequences from a range of sources such as the Emu promoter. However, to properly control gene expression such that the introduced character is most efficiently targeted without other pleiotropic effects, it is generally found that regulatory sequences of plant origin are desirable. Further to this, expression is usually best controlled when a cereal promoter is used in a cereal or a dicot promoter in a dicot.

The most widely used means of cloning such regulatory sequences is by using cDNA libraries from specific tissue types and looking for abundant signals. Naturally, cDNA libraries will not contain the regulatory sequences, hence the abundant cDNA clones must then be used to screen a genomic library so that the regulatory sequences can be identified and cloned.

Methods for Cloning Plant Traits

Cloning Based on Knowledge of Protein Structure/sequence: Many of the first plant traits to be cloned were those in which some degree of biochemical understanding preceded any work with DNA. The understanding may have come from plant systems or from outside the plant kingdom. Most of the cloned genes in these cases were enzymes which were part of a well-characterized biochemical pathway or had a protein structure which was otherwise well known, such as seed storage proteins. Cloning genes based on a knowledge of the protein sequence can be achieved by working backwards from the gene product (amino acid sequence) to the coding sequence or the DNA template. Such an approach has been used for cloning many seed storage proteins, such as the 2S sulphur- rich proteins from Brazil nut. Genes of plant origin have been used in transgenic plants to improve resistance to insect pests. One of the first to be characterized was the cowpea trypsin inhibitor, a protease inhibitor which acts as an antifeedant to a range of insect pests. The protein is naturally produced in seeds of some genotypes of cowpea (*Vigna unguiculata*), which was shown to confer lepidopteran insect resistance when expressed constitutively in tobacco. Another such report from the same group at Durham

University described the use of a snowdrop lectin gene to confer aphid resistance in transgenic tobacco, the first report of genetically engineered resistance to a sucking insect in plants. The lectin from snowdrop (*Galanthus nivalis*), known as GNA, has been demonstrated to be effective against two major rice pests, the brown planthopper and the green leafhopper, as well as the peach potato aphid. When expressed at 0.1% of total leaf protein in tobacco, GNA is antimetabolic to homopteran pests. These levels of protein were accumulated in leaf tissues when placed under the control of the CaMV 35S promoter. However, Hilder *et al.* (1995) postulate that better control would be achieved with the use of a phloem-specific promoter, such as the sucrose synthase 1 promoter from rice.

Another significant group of genes which have been cloned based on prior knowledge of the end product are the seed storage genes, including both seed storage proteins and lipids. In both cases there was considerable biochemical knowledge of the pathways involved in biosynthesis and storage. For seed storage proteins, the genes can be cloned based on the knowledge of the amino acid sequence of the mature protein and targeting signal or transit peptide.

Molecular genetic manipulation of vegetable oils is a major area of research in the developed world. While approximately 90% of vegetable oils are used for human consumption, there are emerging markets for industrial applications including lubricants, paints, cosmetics, plasticizers and soaps.

With increasing emphasis on quality and dietary considerations, there is a desire in developed countries to substantially alter the types and balances of fatty acids found in the major oil crops. Six crop species (soybean, oil palm, rapeseed/canola, sunflower, cottonseed and peanut) account for 84% of the world's vegetable oil production, with the major fatty acids produced being palmitic, stearic, linoleic, linolenic and oleic acids.

However, almost 200 different fatty acids are produced by plants, mostly occurring in nondomesticated plant species. Hence, genetic manipulation of the major crop species, either via conventional breeding or genetic engineering, has the potential to modify the fatty acid types produced and stored by seeds such as soybean and canola.

The biochemistry of fatty acid biosynthesis is well established for many of the major pathways, with enzymes identified which control the chain length of fatty acids, and the degree of saturation, or number of double bonds in the fatty acid chain. The oils stored in seeds have been altered by downregulation of particular enzymes using antisense RNA techniques.

The oleic acid content of rapeseed oil has been increased to 83% (from 62% in conventionally bred cultivars) by downregulating the enzyme D12-oleate desaturase using antisense RNA. A similar result has been achieved using co-suppression, which has produced rapeseed oil with 87% oleic acid content.

Genetic engineering approaches have also been used to introduce totally novel fatty acids to major oilseed crops. Petroselinic acid, an isomer of oleic acid, is found in abundance in the spice plant, coriander (*Coriandrum sativum*). This oil has uses in cosmetics and pharmaceuticals, and its oxidation leads to the formation of lauric acid, used in detergents, and adipic acid, which is used in nylon production.

Hence this is a potentially valuable oil which may enable soybean and canola crops – *i.e.* renewable resources – to replace some of the nonrenewable carbon fuels which are currently used for many industrial polymer and lubrication applications. A cDNA clone from coriander encoding an acyl-ACP desaturase has been genetically engineered into tobacco, resulting in the accumulation of petroselinic acid. Oil crops such as rapeseed and soybean could be genetically modified to produce this fatty acid on a commercial scale.

Hence, here is an example whereby a minor crop plant has become a donor of genetic diversity which will enable the commercial production of a very useful array of products for food and industrial purposes.

Cloning Genes Known Only by Phenotype

However, a more challenging task has been to move beyond proteins and pathways about which sufficient biochemistry is known into cloning genes known by phenotype only. By far the majority of genes fit into this category, which includes many of the most important plant traits such as disease resistance, adaptation to abiotic stresses, phenology and flowering behaviour and fertility. Only in the past few years have genes for traits known only by phenotype been cloned. One of the most interesting areas of plant biology is the understanding of plant–pathogen interactions at the molecular level.

The major methods used for cloning genes known only by phenotype are:

- Transposon tagging.
- Map-based or positional cloning.
- T-DNA tagging.

There have been a number of plant genes cloned using all these methods or modifications of them. There are also examples where a combination of methods has been used to expedite the rate of identification of the unknown sequence.

Transposon Tagging

The proposal that specific pieces of DNA are able to autonomously excise from a particular locus and insert, theoretically at random, into another linked or nonlinked locus was first proposed by Barbara McClintock. Transposable elements are now accepted as ubiquitous, and are particularly well characterized in organisms such as *Escherichia coli*, *Drosophila* and maize. In two plant species, maize and the snapdragon (*Antirrhinum majus*), transposable elements are sufficiently well characterized to be used for mutagenesis and gene tagging. The best understood system is the maize *Ac/Ds* (*Activator/ Dissociation*) family of elements. The behaviour of the *Ac* element is such that it can excise autonomously from its locus, under the control of its own transposase, and insert into another chromosomal position.

Where the *Ac* element is in or near a gene, that gene is inactivated (*i.e.* not expressed), and will appear phenotypically as a recessive mutation. When *Ac* transposes (excises and reinserts at another locus), the gene is commonly returned to normal function, and hence is termed a revertant.

Should this *Ac* element then insert into another gene (transcribed or regulatory sequence), then a new recessive mutation will manifest. Transposition can occur either germinally, which results in an altered phenotype for the whole plant, or somatically, which results in a mosaic mutation, and can often be detected as a sectoring of plant or seed colour. The *Ds* element is actually a group of elements derived from defective *Ac* elements, with a common feature being their lack of effective *Ac* transposase.

Hence these elements are nonautonomous, and will only transpose in the presence of an *Ac* element capable of producing transposase. Equipped with sequence and map information of an *Ac* element in maize, geneticists have been able to clone important maize genes using the transposon-tagging approach.

Table 2: Plant Transposable Elements which have been used for Transposon-Taggine

Element	Species	Type of Element	Size
Ac	Maize	Autonomous	4.6
Ds	Maize	Nonautonomous	0.5–30
Tam1	Snapdragon	Autonomous	15
Tam2	Snapdragon	Nonautonomous	5
En/Spm	Maize	Autonomous	8.3

When a gene is inactivated by an *Ac* element, proof of which can be demonstrated by a low frequency of trait reversion, the *Ac* element can be used as a probe for 'fishing-out' the tagged gene. A tagged gene can be cloned by performing Southern blot analysis with the *Ac* element used as probe, and any polymorphism represents an excision/insertion event. The technique is described more fully in the review by Walbot.

The *Ac* system has proven extremely useful for cloning maize genes for traits such as plant pigment genes (*bronze*; the anthocyanin pathway), starch biosynthesis (*opaque2*), and abscisic acid (AbA) insensitivity. Other maize transposable elements such as *Mutator* (*Mu*) and the *En/Spm* system have been used to clone maize genes, as has the *Tam* system in snapdragon. As a result, the search began for transposable element systems in other species so that a similar approach could be taken.

However, it was soon demonstrated that the well-characterized *Ac/Ds* system could be genetically engineered into other plant species including tobacco, *Arabidopsis*, tomato and petunia. It was subsequently shown that the *Ac* element acted in a similar manner in these heterologous species, undergoing germinal and somatic excision and insertion, and maintaining the propensity to reinsert at a linked site (usually within 4 cM).

As a result, the frequency of mutation in the surrounding area approaches 1024. It was not until 1993 that a plant gene was cloned using the heterologous system, that of the *Ph6* gene in petunia, which is involved in floral pigmentation. Using the transposon-tagging approach, a number of plant disease resistance genes (R genes) have now been cloned. The first of these was the tomato *Cf-9* gene, which confers resistance to race 9 of the fungal pathogen *Cladosporium fulvum*, causal agent of leaf mould. The gene was tagged using a modified *Ac/Ds* system.

The system relied on the use of a stabilized *Ac* element (s*Ac*), which was effectively a disabled *Ac* element that nevertheless produced *Ac* transposase and hence enables *Ds* transposition. The two elements are maintained in separate lines, and hence are both stable until controlled crossing is performed.

This gives a significant advantage in that *Ds* insertion mutants can be maintained indefinitely in the homozygous or hemizygous form, and revertants can be isolated by crossing lines with s*Ac* tomatoes. The transposon-tagging approach used to isolate the *Cf-9* gene, which was known by phenotype only, did have considerable advantages in that much was known of the pathogen.

The hypersensitive response elicited by the virulent strain fitted into the gene-for-gene hypersensitive response proposed by Flor, with a dominant avirulence gene (*Avr9*) interacting with the *Cf-9* gene to form an incompatible or hypersensitive response. *Avr9* had previously been cloned and fully characterized, and was known to specify a 28 amino acid peptide which alone could elicit a necrotic response in resistant tomatoes.

Crosses were made to develop a transgenic tomato line homozygous for *Cf-9*, but heterozygous for *Ds*. Crossing was performed as outlined in Figure. As can be seen, most seedlings died from systemic hypersensitive response, with *Avr9* and *Cf-9* present in all cells. The only survivors would be individuals where *Ds* had inserted into the *Cf-9* gene. Where this happened in plants carrying s*Ac*, the seedlings were variegated for the hypersensitive response (somatic excision), and in the absence of s*Ac*, the survivors were stable (germinal excision).

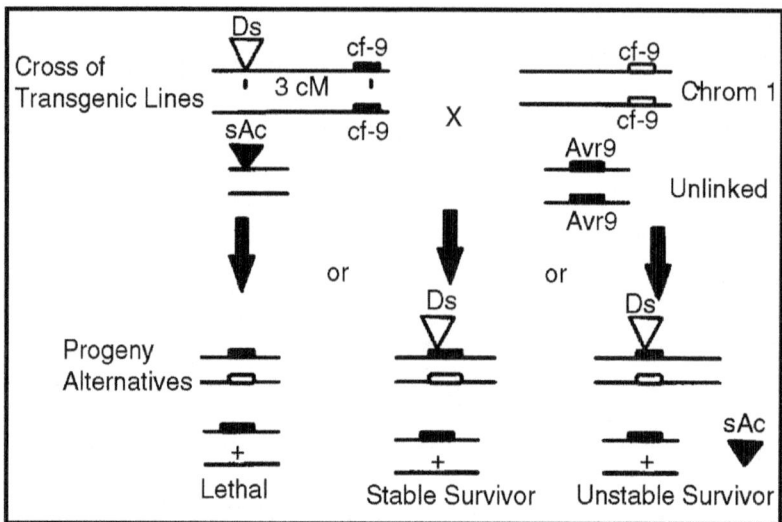

Figure 8: Graphical Representation of the Crosing Strategy of Transgenic Ac/Ds Tomato Lines.

This work was a major undertaking, involving the germination of approximately 160 000 progeny, yielding 118 survivors, representing 63 independent mutations. After considerable effort to characterize these, the *Cf-9* gene was cloned and characterized, with an open reading frame encoding an 863 amino acid protein.

Map-based Cloning

The previously described transposon-tagging scheme utilized to clone *Cf- 9* had an element of the map-based approach by utilizing the phenomenon of *Ac/Ds* transposition to linked sites, with a *Ds* element within 3 cM of the locus of interest, which greatly enhanced the likelihood of insertional mutagenesis. However, map-based cloning requires a much tighter genetic linkage (in centimorgans or recombination fraction), which in turn needs to be converted into a physical distance (kbp of DNA sequence).

The map-based, or positional cloning concept relies on building a well-saturated genetic map of DNA sequences of the organism in question. The genetic map is built up in a polymorphic mapping population, commonly an F2, backcross (BC) or recombinant inbred, with markers such as RFLPs (restriction fragment length polymorphisms). The first RFLP map to be made was of human. Based on such maps, important human genes responsible for disorders such as cystic fibrosis and muscular dystrophy have been cloned in this manner.

Genetic maps of plant species have been made with DNA markers in virtually all important plant species, particularly the major cereals, grain legumes, oilseeds and important vegetable crops including tomato and potato. The RFLP marker system is based on Southern blot analysis, while many of the newer technologies use the polymerase chain reaction (PCR).

These include random amplified polymorphic DNA (RAPDs), amplified fragment length polymorphisms (AFLPs), and a number of different marker systems based on simple sequence repeats (SSRs), which may use PCR, such as inter simple sequence repeat (ISSR)-PCR, or Southern analysis. The applications of DNA markers to plant genetic analysis and breeding are many, and the subject has been reviewed by various authors.

The population for saturation mapping should be segregating in a Mendelian manner for the trait of interest. This requires a lot of tedious but demanding work, and it may be that particular populations such as near isogenic lines, or DNA pooling strategies, such as bulked segregant analysis, will allow the mapping effort to be more targeted to the region of interest, hence improving the rate of map saturation.

It is desirable to obtain extremely close genetic linkage to the locus of interest (<1 cM), with flanking markers if possible. It can be of considerable advantage to identify a marker which co-segregates (say within 0.1 cM) with the trait. The next step is to convert the genetic map into a physical map of the chromosome region, translating distances from centimorgans to nucleotides.

The Gene of Interest (in this illustration,R) is Flanked by Markers such as RFLPs or SSRs. A Series of Overlapping YAC Contigs is used to cover th Chromosomal Region Containing the Gene.The YAC Which Contains the Complete Gene is Marked in Bold.This is then Subcloned or Matched to a cDNA Library to Idenfify the Gene(s) Contained the Region, and Ultimately Identify the R Gene.

Physical mapping then involves the use of YAC or BAC clones which span the region of interest. These YAC/BAC contigs will span the region between two flanking DNA marker sequences as shown in Figure, a process known as chromosome walking. In this example, two YAC/BAC clones contain the gene of interest. Methodology is being developed to enable BAC clones to be directly transformed into plants, to find which one contains the gene of interest (in the example, the *R* gene). Alternatively, the YAC/BAC must be subcloned (for example as a cDNA library), and these subclones can be put into the susceptible plant and tested for the R phenotype.

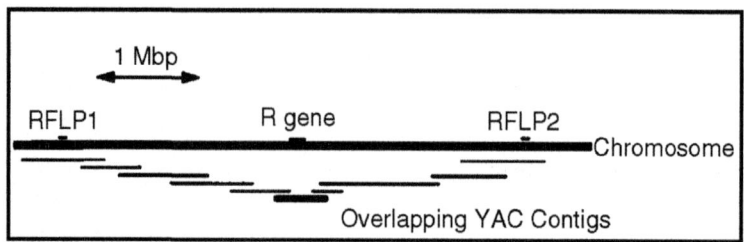

Figure 9: Generalized Strategy of Chromosome Walking Used in Map-based gene Cloning.

The first R gene to be cloned in plants via map-based cloning was the *Pto* gene, which confers resistance to *Pseudomonas syringae* pv. *lycopersici*, the ausal agent of bacterial speck in tomato.

The starting points for the work were:
- A high-density RFLP map of tomato.
- A tomato YAC library.

The *Pto* locus had been previously mapped to tomato chromosome 5 and one RFLP marker co-segregated with *Pto*. Hence this was a useful starting point to screen the YAC library. A 400 kb YAC was identified, and via inverse PCR, some end-specific probes were generated. The left arm of the YAC was 1.8 cM from *Pto*, with the right arm perfectly co-segregating. This YAC was then used to screen a 920 000 plaque leaf cDNA library, from which 30 were selected. One of these was shown to co-segregate with *Pto*. This cDNA was then transformed into a susceptible tomato cultivar under the control of the CaMV 35S promoter, where it conferred resistance. Hence the gene was identified, and could then be sequenced and characterized.

The success of this approach was partly due to the small genome size of tomato (950 Mbp), and a relatively low level of repeat sequence. Such an approach is much moredifficult in species with larger genomes such as maize (2300–2700 Mbp) and wheat (16 000 Mbp). One such example of this problem is the 79 000 clone maize YAC library produced by Zeneca Seeds. As reported by K. Edwards, only 15% of the YACs have unique ends, with the remainder terminating in repeat sequences, which makes chromosome walking very difficult. The synteny of genomes is a major advantage in this instance. It is now well established that there is a high degree of synteny of genome organization, as seenin the co-linearity of maps of the most important grass genomes. This has resulted in the great interest in using this co-linearity to facilitate cloning of genes in large genomes such as maize and wheat, by starting with the smaller genomes, particularly rice, and to a lesser extent, sorghum.

One particular study of maize YACs demonstrated that a single YAC with the *Adh1* sequence contained 36 different repetitive sequences, whereas these repetitive sequences were largely absent in corresponding areas of the rice and sorghum genomes, a pattern seen also for the *a1–sh2* region in all three genomes. In a similar manner, many genes of interest for *Brassica* improvement are being identified in the smaller *Arabidopsis thaliana* genome.

Paterson *et al.* have recently demonstrated the potential of this approach, by identifying quantitative trait loci (QTLs) for seed mass and phenoloy which were co-linear on molecular marker maps of maize, sorghum and rice. They proposed that, as the maize genome is four and six times larger than the sorghum and rice genomes, respectively, the genes of interest in maize may be cloned in sorghum/ rice more expediently. It must be remembered that QTLs are subject to genotype 3 environment interaction, which will be a major complicating factor in attempts to clone and characterize a QTL.

T-DNA Tagging

When T-DNA is inserted into the plant genome, it does so via illegitimate recombination. In the same way as a transposon, the insertion of T-DNA into a gene will inactivate it, which phenotypically will appear as a mutation. The comparative advantage of such a method is that stable mutations will be conferred. This can be a disadvantage, however, as reversions, such as those made possible by the *Ac/Ds* system are not possible. It is also the case that each independent transformation event will result in just one mutation event (or more if multiple independent insertions are made), whereas the heterologous transposon approach allows multiple mutation events to take place from a single initial transformant. Hence, mutated genes with T-DNA insertions can be cloned, in much the same manner as those with transposon insertions, by using part of the T-DNA as a probe in Southern blots.

The procedure for selecting out the tagged gene is quite straightforward, particularly when the plant contains a single T-DNA copy. Howeer, this means that a large number of independent transformants must be generated. This can be achieved with species such as *Arabidopsis*, where Feldmann demonstrated that a seed-based transformation system using *Agrobacterium* can be quite efficient. Feldmann demonstrated that by inoculating approximately 1000 seeds, approximately 300 000 selfed T2 seed can be collected. They estimated that approximately 8000 transformants could be recovered from an experiment of this scale and, using this approach, cloned a gene involved in trichome development and a dwarf gene from *Arabidopsis*.

A regulatory gene in the ethylene biosynthesis pathway of *Arabidopsis* has aso been cloned using the T-DNA tagging strategy. With modification, T-DNA tagging methodology can be used to clone plant promoter sequences. Koncz *et al.* designed a vector which contained a promoterless antibiotic selection gene (aminoglycoside phosphotransferase II) and transformed this into *Nicotiana* and *Arabidopsis*. In this way, promoters could be 'trapped' as plants that were regenerated on selective antibiotic were antibiotic resistant because the T-DNA was inserted downstream of a promoter such that the gene was expressed.

The most serious limitation to the T-DNA tagging approach is that large numbers of independent transformation events must be generated, which is not currently possible with most species. This is really only feasible with a system such as the seed transformation technique developed for *Arabidopsis*, or possibly the 'whole-leaf ' transformation system developed for tobacco.

Chapter 6

Life's Genetic Tree

The millions of different living things on Earth—plants, bacteria, insects, chimps, people, and everything else—all came to be because of a process called biological evolution, in which organisms change over time. Because of biological evolution, early humans gained the ability to walk on two feet. Because of evolution, air-breathing whales can live in the ocean despite being mammals like us. Because of evolution, some bacteria can live in scalding water, others can survive in solid ice, and still others can live deep in the Earth eating only rocks!

Evolution happens every day, and it affects every species—including us. It changes entire populations, not individuals. And it has a big impact on medical research.

- Everything Evolves
- Selective Study
- Clues from Variation
- Living Laboratories
- The Genome Zoo
- Starting All Over Again
- Genes Meet Environment
- Genetics and You: You've Got Rhythm!
- Animals Helping People
- My Collaborator Is a Computer
- The Tools of Genetics: Unlimited DNA
- Got It?

Everything Evolves

To understand evolution, let's go back in time a century and a half to 1854, when the British naturalist Charles Darwin published

The Origin of Species, a book that proposed an explanation for how evolution works. The main concept in evolution is that all living things share a common ancestor. The very earliest ancestor of all life forms on Earth lived about 4 billion years ago. From that early organism, millions of types of creatures—some living and some now extinct—have evolved. Evolution requires diversity. You can tell that living things are diverse just by walking down the street and looking around you. Individual people are very different from one another. Chihuahuas are different from Great Danes, and Siamese cats are different from tabbies.

Evolution also depends on inheritance. Many of our unique characteristics are inherited—they are passed from parent to offspring. This is easy to see: Dalmatian puppies look like Dalmatians, not Chihuahuas. Petunias grow differently from pansies. Evolution works *only* on traits that are inherited. Finally, as you probably already know, evolution favours the "fittest." Through a process called natural selection, only some offspring within a given generation will survive long enough to reproduce. As an example, consider houseflies, each of which lays thousands of eggs every year. Why haven't they taken over the world? Because almost all of the baby houseflies die. The flies that survive are the ones that can find something to eat and drink... the ones that avoid being eaten, stepped on, or swatted... and the ones that don't freeze, drown, or land on a bug zapper.

The flies that survive all these ways to die have what it takes to outlive most of their brothers and sisters. These inherited traits give an organism a survival edge. Those who survive will mate with each other and will pass on to the next generation some of their DNA that encoded these advantageous traits. Of course, not all aspects of survival are determined by genes. Whether a fly gets swatted depends on genes that affect its reflexes—whether it's fast enough to avoid the swatter—but also on the environment. If there's no human around waving the swatter, the fly is quite likely to survive, regardless of its reflexes.

Evolution often takes a long time to make a difference. But it can also happen very quickly, especially in organisms with short lifespans. For example, as you read earlier, some bacteria have molecular features that let them survive in the presence of antibiotics.When you take an antibiotic medicine, antibiotic-resistant bacteria flourish while antibiotic-sensitive bacteria die. Because antibiotic resistance is a growing public health threat, it's important to take the whole dosage of antibiotic medicine, not stop when you feel better. And you should take antibiotics only when they're needed, not for colds or other viral infections, which antibiotics can't treat. Viruses must simply run their course.

Selective Study

Scientists doing medical research are very interested in genetic variants that have been selected by evolution. For example, researchers recently discovered a rare genetic variant that protects people from getting AIDS. A genetic variant is a different version of a gene, one that has a slightly different sequence of nucleotides. Scientists think that the rare variant of a gene called CCR5 originally may have been selected during evolution because it made people resistant to an organism unrelated to HIV.

Montgomery Slatkin of the University of California, Berkeley, has used mathematical modeling techniques to show that natural selection over time could explain the frequency of the CCR5 variant in human populations. The work indicates that the CCR5 gene variant's ability to protect against AIDS may contribute to keeping it in the human gene pool.

So, through evolution, living things change. Sometimes, that's good for us, as when humans understand HIV resistance in hopes of preventing AIDS. But sometimes the changes aren't so great —from a human perspective, anyway—as when bacteria become resistant to antibiotics.

Whether the consequences of evolutionary change are good or bad, understanding the process can help us develop new strategies for fighting disease.

Clues from Variation

Scientists know quite a bit about how cells reshuffle genetic information to create each person's unique genome. But many details are missing about how this genetic variation contributes to disease, making for a very active area of research.

Different nucleotides (in this example, A or G) can appear in the DNA sequence of the same chromosome from two different individuals, creating a single-nucleotide polymorphism (SNP).

What scientists do know is that most of the human genome is the same in all of us. A little bit of genetic variation—differences that account for much less than 1 percent of our DNA—gives each of us a unique personality, appearance, and health profile.

The parts of the human genome where the DNA sequences of many individuals vary by a single nucleotide are known as single-nucleotide polymorphisms (abbreviated SNPs and pronounced "snips"). For example, let's say that a certain nucleotide in one of your genes is A. In your uncle, however, the nucleotide in the same place on the same gene might be G. You and your uncle have slightly different versions of that gene. Scientists call the different gene versions alleles.

Haplotypes are combinations of gene variants, or SNPs, that are likely to be inherited together within the same chromosomal region. In this example, an original haplotype (top) evolved over time to create three newer haplotypes that each differ by a few nucleotides (red).

If two genes sit right next to each other on a chromosome, the SNPs in those genes tend to be inherited together. This set of neighboring SNPs is called a haplotype. Most chromosome regions have only a few, common haplotypes among all humans. As it turns out, these few haplotypes—in different combinations in each person—appear to account for most of the variation from person to person in a population.

Scientists can use haplotype information to compare the genes of people affected by a disease with those of unaffected people. For example, this approach revealed a genetic variation that substantially increases the risk of age-related macular degeneration, the leading cause of severe vision loss in the elderly. Scientists discovered that a single SNP—one nucleotide in the 3 billion-nucleotide human genome—makes some people more likely to get this eye disease. The discovery paves the way for better diagnostic tests and treatments.

What about other diseases? In October 2005, an international scientific team published a catalogue of human haplotypes. Researchers are looking through the catalogue in an effort to identify genes that determine susceptibility to many common diseases, including asthma, diabetes, cancer, and heart disease.

But not all SNPs are in genes. Scientists studying genetic variation have also found SNPs in DNA that doesn't encode proteins. Nonetheless, some of these SNPs appear to affect gene activity.

Some researchers suspect that the "cryptic" (hidden) variation associated with SNPs in non-coding DNA plays an important role in determining the physical characteristics and behaviours of an organism.

Loren Rieseberg of Indiana University in Bloomington is one scientist who would love to take the mystery out of cryptic variation. He wants to know how this non-coding genetic variation can help organisms adapt to new environments. He's also curious about whether it can create problems for some individuals.

You might be surprised to learn that Rieseberg's principal research subject is the sun- flower. Although many plants produce only one generation a year, plants like sunflowers can be very useful tools for researchers asking fundamental questions about genetics.

Because their genetic material is more malleable than that of many animals, plants are excellent models for studying how evolution works. Wild sunflowers appealed to Rieseberg because there are several species that live in different habitats. Two ancient species of wild sunflowers grow in moderate climates and are broadly distributed throughout the central and western United States. Three recently evolved sunflower species live in more specialized environments: One of the new species grows on sand dunes, another grows in dry desert soil, and the third species grows in a salt marsh.

To see how quickly new plant species could evolve, Rieseberg forced the two ancient sunflowers to interbreed with each other, something plants but not other organisms can do. Among the hybrid progeny were sunflowers that were just like the three recently evolved species! What that means is that Rieseberg had stimulated evolution in his lab, similar to what actually happened in nature some 60,000 to 200,000 years ago, when the newer species first arose.

That Rieseberg could do this is pretty amazing, but the really interesting part is how it happened. Scientists generally assume that, for a new species with very different characteristics to evolve, a lot of new mutations have to occur. But when Rieseberg looked at the genomes of his hybrid sunflowers, he was surprised to find that they were just cut-and-pasted versions of the ancient sunflower species' genomes: large chunks had been moved rather than many new SNPs created.

Rieseberg reasons that plants stash away unused genetic material, giving them a ready supply of ingredients they can use to adapt quickly to a new environment. It may be that human genomes can recycle unused genetic material to confront new challenges, as well.

Living Laboratories

The Genome Zoo

Scientists often use an image of a tree to depict how all organisms, living and extinct, are related to a common ancestor. In this "tree of life," each branch represents a species, and the forks between branches show when the species represented by those branches became different from one another. For example, researchers estimate that the common ancestor of humans and chimpanzees lived about 6 million years ago.

While it is obvious just by looking that people have a lot in common with our closest living relatives, chimpanzees, what about more distant species? If you look at an evolutionary tree, you'll see that humans are related to mice, worms, and even bacteria. The ancestral species that gave rise to both humans and bacteria was alive a lot longer ago than the ancestor of humans and chimpanzees, yet we still share hundreds of genes with bacteria. Scientists use the term comparative genomics to describe what they're doing when they compare the genomes of different species to see how similar (or how different!) the species' DNA sequences are. Sequences that the species have in common are the molecular footprints of an ancestor of those species.

Why are "old" DNA sequences still in our genomes? It turns out that nature is quite economical, so DNA sequences that are responsible for something as complicated and important as controlling gene activity may stay intact for millions of years.

Comparative genomic studies also have medical implications. What would you do if you wanted to develop new methods of preventing, diagnosing, or treating a human disease that animals don't get? If people have a gene that influences their risk for a disease, and mice have the gene too, you could study some aspect of the disease in mice, even though they don't ever have the symptoms of the disease. You could even study the disease in yeast, if it has the gene, as well.

Starting all over Again

Stem cells—what embryos are made up of just days after an egg is fertilized by a sperm—have the amazing ability to grow up into any kind of cell: skin, heart, muscle, nerve, and everything else. How do they do it? Intrigued by the potential of these masterful cells, researchers want to know just what it is that gives stem cells their ability to change into anything, upon the body's request, but stay in the "I can do anything" state until asked.

Some basic researchers are trying to figure out how stem cells work by using a unique model system: tiny, freshwater worms called planarians. These worms are like stem cells in the sense that they can regenerate. You can cut up planarians into hundreds of pieces, and each piece will regenerate into an intact worm that looks the same as all the others.

Planarians' resemblance to stem cells isn't just coincidence. Scientists have discovered that planarians can perform the amazing act of regeneration due to the presence of, yes, specialized stem cells in their bodies.

Developmental biologist Alejandro Sánchez Alvarado of the University of Utah School of Medicine in Salt Lake City used the gene-silencing technique RNAi (see *RNA Interference*) to search for planarian genes that were essential for regeneration. He found 240 genes that, when silenced, caused a physical defect in the worm's growth and regenerative ability. Interestingly, 16 percent of these looked very much like genes that had been linked to human disease!

Sánchez Alvarado and his team hope to figure out how regeneration genes allow the specialized stem cells within the worm to travel to a wounded site and "turn into" any of the 30 or so cell types needed to recreate a mature worm.

Although humans are only distantly related to planarians, we have many of the same genes, so these findings have the potential to reveal strategies for regenerating injured body parts in people, too.

Genes Meet Environment

If toxins from the environment get into our bodies, they don't always make us sick. That's because liver enzymes come to our rescue to make the chemicals less harmful. The genes that encode those enzymes are under constant evolutionary pressure to adapt quickly to new toxins. For example, certain liver enzymes called cytochrome P450 proteins metabolize, or break down, hormones that our bodies make as well as many of the foreign substances that we encounter. These include harmful molecules like cancer-causing agents as well as beneficial ones, like medicines. In fact, just two genes within the cytochrome P450 family, abbreviated 3A4 and 3A5, encode proteins that process more than half of all of the medicines that are sold today.

Since the chemicals to which people are exposed vary so widely, a scientist might predict that there would be different variants of cytochrome P450 genes in different human populations. Using comparative genomics, researchers such as Anna Di Rienzo of the University of Chicago have shown that this is indeed the case. Di Rienzo has found many sequence differences within these genes in people living throughout the world.

It turns out that one variant of the gene that encodes the cytochrome P450 3A5 protein makes this enzyme very efficient at breaking down cortisol, a hormone that raises salt levels in the kidneys and helps the body retain water. Di Rienzo compared the DNA sequences of the 3A5 gene in DNA samples taken from more than 1,000 people representing over 50 populations worldwide. She was amazed to find a striking link between the existence of the gene variant and the geographic locale of the people who have it.

Di Rienzo discovered that African populations living very close to the equator were more likely than other populations to have the salt-saving version of the 3A5 gene. She suggests that this is because this gene variant provides a health advantage for people living in a very hot climate, since retaining salt helps ward off dehydration caused by intense heat.

However, there seems to be a cost associated with that benefit— the 3A5 gene variant raises the risk for some types of high blood pressure. That means that in environments in which retaining salt is not beneficial, evolution selects against this gene variant.

Another scientist who studies interactions between genes and the environment is Serrine Lau of the University of Arizona in Tucson. She studies a class of harmful molecules called polyphenols, present in cigarette smoke and car exhaust, that cause kidney cancer in rats, and perhaps, in people. Lau discovered that rats and humans who are more sensitive to some of the breakdown products of polyphenols have an unusual DNA sequence—a genetic signature—that increases their risk of developing cancer. She suspects that the gene that is affected encodes a tumor suppressor: a protein that prevents cancer from developing.

In people and rats with the genetic signature, she reasons, the tumor suppressor doesn't work right, so tumors grow. Taking this logic one step further, it may be that certain people's genetic make-up makes them unusually susceptible to DNA damage caused by exposure to carcinogens. If doctors could identify those at risk, Lau says, such people could be forewarned to avoid contact with specific chemicals to protect their health. However, think about this scenario: Who should make those decisions? For example, would it be ethical for an employer to refuse to hire somebody because the person has a genetic signature that makes him or her more likely to get cancer if exposed to a chemical used in the workplace? Tough question.

Plant Breeding

Plant breeding is the art and science of changing the genetics of plants for the benefit of humankind. Plant breeding can be accomplished through many different techniques ranging from simply selecting plants with desirable characteristics for propagation, to more complex molecular techniques. Plant breeding has been practiced for thousands of years, since near the beginning of human civilization. It is now practiced worldwide by individuals such as gardeners and farmers, or by professional plant breeders employed by organizations such as government institutions, universities, crop-specific industry associations or research centres.

International development agencies believe that breeding new crops is important for ensuring food security by developing new varieties that are higher-yielding, resistant to pests and diseases, drought-resistant or regionally adapted to different environments and growing conditions.

Domestication

This map shows the sites of domestication for a number of crops. Places where crops were initially domesticated are called centres of origin.

Plant breeding in certain situations may lead the domestication of wild plants. Domestication of plants is an artificial selection process conducted by humans to produce plants that have more desirable traits than wild plants, and which renders them dependent on artificial (usually enhanced) environments for their continued existence. The practice is estimated to date back 9,000-11,000 years. Many crops in present day cultivation are the result of domestication in ancient times, about 5,000 years ago in the Old World and 3,000 years ago in the New World. In the Neolithic period, domestication took a minimum of 1,000 years and a maximum of 7,000 years. Today, all of our principal food crops come from domesticated varieties. Almost all the domesticated plants used today for food and agriculture were domesticated in the centres of origin. In these centres there is still a great diversity of closely related wild plants, the so called crop wild relatives that can also be used for improving modern cultivars by plant breeding.

A plant whose origin or selection is due primarily to intentional human activity is called a cultigen, and a cultivated crop species that has evolved from wild populations due to selective pressures from traditional farmers is called a landrace. Landraces, which can be the result of natural forces or domestication, are plants (or animals) that are ideally suited to a particular region or environment. An example are the landraces of rice, *Oryza sativa* subspecies *indica*, which was developed in South Asia, and *Oryza sativa* subspecies *japonica*, which was developed in China.

Classical Plant Breeding

Classical plant breeding uses deliberate interbreeding (crossing) of closely or distantly related individuals to produce new crop varieties or lines with desirable properties. Plants are crossbred to introduce traits/genes from one variety or line into a new genetic background. For example, a mildew-resistant pea may be crossed with a high-yielding but susceptible pea, the goal of the cross being to introduce mildew resistance without losing the high-yield characteristics. Progeny from the cross would then be crossed with the high-yielding parent to ensure that the progeny were most like the high-yielding parent, (backcrossing). The progeny from that cross would then be tested for yield and mildew resistance and high-yielding resistant plants would be further developed. Plants may also be crossed with themselves to produce inbred varieties for breeding.

Classical breeding relies largely on homologous recombination between chromosomes to generate genetic diversity. The classical plant breeder may also makes use of a number of *in vitro* techniques such as protoplast fusion, embryo rescue or mutagenesis to generate diversity and produce hybrid plants that would not exist in nature. Traits that breeders have tried to incorporate into crop plants in the last 100 years include:

1. Increased quality and yield of the crop
2. Increased tolerance of environmental pressures (salinity, extreme temperature, drought)

3. Resistance to viruses, fungi and bacteria
4. Increased tolerance to insect pests
5. Increased tolerance of herbicides.

Before World War II

Intraspecific hybridization within a plant species was demonstrated by Charles Darwin and Gregor Mendel, and was further developed by geneticists and plant breeders. In the United Kingdom in the 1880s was the pioneering work of Gartons Agricultural Plant Breeders. In the early 20th century, plant breeders realized that Mendel's findings on the non-random nature of inheritance could be applied to seedling populations produced through deliberate pollinations to predict the frequencies of different types.

In 1908, George Harrison Shull described heterosis, also known as hybrid vigor. Heterosis describes the tendency of the progeny of a specific cross to outperform both parents. The detection of the usefulness of heterosis for plant breeding has led to the development of inbred lines that reveal a heterotic yield advantage when they are crossed. Maize was the first species where heterosis was widely used to produce hybrids.

By the 1920s, statistical methods were developed to analyse gene action and distinguish heritable variation from variation caused by environment. In 1933, another important breeding technique, cytoplasmic male sterility (CMS), developed in maize, was described by Marcus Morton Rhoades. CMS is a maternally inherited trait that makes the plant produce sterile pollen. This enables the production of hybrids without the need for labour intensive detasseling. These early breeding techniques resulted in large yield increase in the United States in the early 20th century. Similar yield increases were not produced elsewhere until after World War II, the Green Revolution increased crop production in the developing world in the 1960s.

After World War II

Following World War II a number of techniques were developed that allowed plant breeders to hybridize distantly related species, and artificially induce genetic diversity. When distantly related species are crossed, plant breeders make use of a number of plant tissue culture techniques to produce progeny from otherwise fruitless mating. Interspecific and intergeneric hybrids are produced from a cross of related species or genera that do not normally sexually reproduce with each other. These crosses are referred to as *Wide crosses*. For example, the cereal triticale is a wheat and rye hybrid. The cells in the plants derived from the first generation created from the cross contained an uneven number of chromosomes and as result was sterile.

The cell division inhibitor colchicine was used to double the number of chromosomes in the cell and thus allow the production of a fertile line. Failure to produce a hybrid may be due to pre- or post-fertilization incompatibility. If fertilization is possible between two species or genera, the hybrid embryo may abort before maturation. If this does occur the embryo resulting from an interspecific or intergeneric cross can sometimes be rescued and cultured to produce a whole plant. Such a method is referred to as *Embryo Rescue*. This technique has been used to produce new rice for Africa, an interspecific cross of Asian rice *(Oryza sativa)* and African rice *(Oryza glaberrima)*.

Hybrids may also be produced by a technique called protoplast fusion. In this case protoplasts are fused, usually in an electric field. Viable recombinants can be regenerated in culture.

Chemical mutagens like EMS and DMS, radiation and transposons are used to generate mutants with desirable traits to be bred with other cultivars- a process known as *Mutation Breeding*. Classical plant breeders also generate genetic diversity within a species by exploiting a process called somaclonal variation, which occurs in plants produced from tissue culture, particularly plants derived from callus. Induced polyploidy, and the addition or removal of chromosomes using a technique called chromosome engineering may also be used.

When a desirable trait has been bred into a species, a number of crosses to the favoured parent are made to make the new plant as similar to the favoured parent as possible. Returning to the example of the mildew resistant pea being crossed with a high-yielding but susceptible pea, to make the mildew resistant progeny of the cross most like the high-yielding parent, the progeny will be crossed back to that parent for several generations. This process removes most of the genetic contribution of the mildew resistant parent. Classical breeding is therefore a cyclical process.

It should be noted that with classical breeding techniques, the breeder does not know exactly what genes have been introduced to the new cultivars. Some scientists therefore argue that plants produced by classical breeding methods should undergo the same safety testing regime as genetically modified plants.

There have been instances where plants bred using classical techniques have been unsuitable for human consumption, for example the poison solanine was unintentionally increased to unacceptable levels in certain varieties of potato through plant breeding. New potato varieties are often screened for solanine levels before reaching the marketplace.

Modern Plant Breeding

Modern plant breeding uses techniques of molecular biology to select, or in the case of genetic modification, to insert, desirable traits into plants.

Steps of Plant Breeding

The following are the major activities of plant breeding;

1. Creation variation
2. Selection
3. Evaluation
4. Release
5. Multiplication
6. Distribution of the new variety.

Marker Assisted Selection

Sometimes many different genes can influence a desirable trait in plant breeding. The use of tools such as molecular markers or DNA fingerprinting can map thousands of genes. This allows plant breeders to screen large populations of plants for those that possess the trait of interest. The screening is based on the presence or absence of a certain gene as determined by laboratory procedures, rather than on the visual identification of the expressed trait in the plant.

Reverse Breeding and Doubled Haploids (DH)

A method for efficiently producing homozygous plants from a heterozygous starting plant, which has all desirable traits. This starting plant is induced to produce doubled haploid from haploid cells, and later on creating homozygous/doubled haploid plants from those cells. While in natural offspring recombination occurs and traits can be unlinked from each other, in doubled haploid cells and in the resulting DH plants recombination is no longer an issue. There, a recombination between two corresponding chromosomes does not lead to un-linkage of alleles or traits, since it just leads to recombination with its identical copy. Thus, traits on one chromosome stay linked. Selecting those offspring having the desired set of chromosomes and crossing them will result in a final F1 hybrid plant, having exactly the same set of chromosomes, genes and traits as the starting hybrid plant. The homozygous parental lines can reconstitute the original heterozygous plant by crossing, if desired even in a large quantity. An individual heterozygous plant can be converted into a heterozygous variety (F1 hybrid) without the necessity of vegetative propagation but as the result of the cross of two homozygous/doubled haploid lines derived from the originally selected plant. patent

Genetic Modification

Genetic modification of plants is achieved by adding a specific gene or genes to a plant, or by knocking down a gene with RNAi, to produce a desirable phenotype. The plants resulting from adding a gene are often referred to as transgenic plants. If for genetic modification genes of the species or of a crossable plant are used under control of their native promoter, then they are called cisgenic plants. Genetic modification can produce a plant with the desired trait or traits faster than classical breeding because the majority of the plant's genome is not altered.

To genetically modify a plant, a genetic construct must be designed so that the gene to be added or removed will be expressed by the plant. To do this, a promoter to drive transcription and a termination sequence to stop transcription of the new gene, and the gene or genes of interest must be introduced to the plant. A marker for the selection of transformed plants is also included.

In the laboratory, antibiotic resistance is a commonly used marker: plants that have been successfully transformed will grow on media containing antibiotics; plants that have not been transformed will die. In some instances markers for selection are removed by backcrossing with the parent plant prior to commercial release.

The construct can be inserted in the plant genome by genetic recombination using the bacteria *Agrobacterium tumefaciens* or *A. rhizogenes*, or by direct methods like the gene gun or microinjection. Using plant viruses to insert genetic constructs into plants is also a possibility, but the technique is limited by the host range of the virus. For example, Cauliflower mosaic virus (CaMV) only infects cauliflower and related species. Another limitation of viral vectors is that the virus is not usually passed on the progeny, so every plant has to be inoculated.

The majority of commercially released transgenic plants, are currently limited to plants that have introduced resistance to insect pests and herbicides. Insect resistance is achieved through incorporation of a gene from *Bacillus thuringiensis* (Bt) that encodes a protein that is toxic to some insects.

For example, the cotton bollworm, a common cotton pest, feeds on Bt cotton it will ingest the toxin and die. Herbicides usually work by binding to certain plant enzymes and inhibiting their action. The enzymes that the herbicide inhibits are known as the herbicides *target site*. Herbicide resistance can be engineered into crops by expressing a version of *target site* protein that is not inhibited by the herbicide. This is the method used to produce glyphosate resistant crop plants.

Genetic modification of plants that can produce pharmaceuticals (and industrial chemicals), sometimes called *pharmacrops*, is a rather radical new area of plant breeding.

Issues and Concerns

Modern plant breeding, whether classical or through genetic engineering, comes with issues of concern, particularly with regard to food crops. The question of whether breeding can have a negative effect on nutritional value is central in this respect. Although relatively little direct research in this area has been done, there are scientific indications that, by favouring certain aspects of a plant's development, other aspects may be retarded. A study published in the *Journal of the American College of Nutrition* in 2004, entitled *Changes in USDA Food Composition Data for 43 Garden Crops, 1950 to 1999*, compared nutritional analysis of vegetables done in 1950 and in 1999, and found substantial decreases in six of 13 nutrients measured, including 6% of protein and 38% of riboflavin. Reductions in calcium, phosphorus, iron and ascorbic acid were also found. The study, conducted at the Biochemical Institute, University of Texas at Austin, concluded in summary: *"We suggest that any real declines are generally most easily explained by changes in cultivated varieties between 1950 and 1999, in which there may be trade-offs between yield and nutrient content. "*

The debate surrounding genetically modified food during the 1990s peaked in 1999 in terms of media coverage and risk perception, and continues today- for example, *"Germany has thrown its weight behind a growing European mutiny over genetically modified crops by banning the planting of a widely grown pest-resistant corn variety.".* The debate encompasses the ecological impact of genetically modified plants, the safety of genetically modified food and concepts used for safety evaluation like substantial equivalence.

Such concerns are not new to plant breeding. Most countries have regulatory processes in place to help ensure that new crop varieties entering the marketplace are both safe and meet farmers' needs. Examples include variety registration, seed schemes, regulatory authorizations for GM plants, etc.

Plant breeders' rights is also a major and controversial issue. Today, production of new varieties is dominated by commercial plant breeders, who seek to protect their work and collect royalties through national and international agreements based in intellectual property rights. The range of related issues is complex. In the simplest terms, critics of the increasingly restrictive regulations argue that, through a combination of technical and economic pressures, commercial breeders are reducing biodiversity and significantly constraining individuals (such as farmers) from developing and trading seed on a regional level. Efforts to strengthen breeders' rights, for example, by lengthening periods of variety protection, are ongoing.

When new plant breeds or cultivars are bred, they must be maintained and propagated. Some plants are propagated by asexual means while others are propagated by seeds. Seed propagated cultivars require specific control over seed source and production procedures to maintain the integrity of the plant breeds results. Isolation is necessary to prevent cross contamination with related plants or the mixing of seeds after harvesting. Isolation is normally accomplished by planting distance but in certain crops, plants are enclosed in greenhouses or cages (most commonly used when producing F1 hybrids.)

Participatory Plant Breeding

Participatory Plant Breeding is carried out, for example, in northern Vietnam, where government scientists work with farmers from the Muong people ethnic minority to improve local rice varieties.

The development of agricultural science, with phenomenon like the Green Revolution arising, have left millions of farmers in developing countries, most of whom operate small farms under unstable and difficult growing conditions, in a precarious situation. The adoption of new plant varieties by this group has been hampered by the constraints of poverty and the international policies promoting an industrialised model of agriculture. Their response has been the creation of a novel and promising set of research methods collectively known as participatory plant breeding.

Participatory means that farmers are more involved in the breeding process and breeding goals are defined by farmers instead of international seed companies with their large-scale breeding programs. Farmer's groups and NGOs, for example, may wish to affirm local people's rights over genetic resources, produce seeds themselves, build farmers' technical expertise, or develop new products for niche markets, like organically grown food.

Chapter 7

Quantitative Genetics, Genomics, Medical Genetics and Behavioural Genetics

Quantitative Genetics

Quantitative genetics is the study of continuous traits (such as height or weight) and their underlying mechanisms. It is effectively an extension of simple Mendelian inheritance in that the combined effect of the many underlying genes results in a continuous distribution of phenotypic values.

The field was founded, in evolutionary terms, by the originators of the modern synthesis, R.A. Fisher, Sewall Wright and J. B. S. Haldane, and aimed to predict the response to selection given data on the phenotype and relationships of individuals. Analysis of Quantitative Trait Loci, or QTL, is a more recent addition to the study of quantitative genetics. A QTL is a region in the genome that affects the trait or traits of interest. Quantitative trait loci approaches require accurate phenotypic, pedigree and genotypic data from a large number of individuals.

Traits

Quantitative genetics is not limited to continuous traits, but to all traits that are determined by many genes. This includes:

- Continuous traits are quantitative traits with a continuous phenotypic range. They are often polygenic, and may also be influenced significantly by environmental effects.
- Meristic traits or other ordinal numbers are expressed in whole numbers, such as number of offspring, or number of bristles on a fruit fly. These traits can be either treated as approximately discontinuous traits or as threshold traits.

- Some qualitative traits can be treated as if they have an underlying quantitative basis, expressed as a threshold trait (or multiple thresholds). Some human diseases (such as, schizophrenia) have been studied in this manner.

Basic Principles

The phenotypic value (P) of an individual is the combined effect of the genotypic value (G) and the environmental deviation (E):

$$P = G + E$$

The genotypic value is the combined effect of all the genetic effects, including nuclear genes, mitochondrial genes and interactions between the genes. It is worthwhile to note that the mathematics *is* related to the genetics: for which the brief following revision may be useful. In disomic (diploid) organisms, a nucleus gene is represented twice in the gene-set ("genotype"), one contribution being provided by each parent during sexual reproduction. Each "gene" is located at a particular place (a "locus" - the Latin word for place; plural "loci") on corresponding chromosomes (homologues), one from each parent.

Any gene may have several functional forms in the species as a whole, and each of these may lead to outwardly different "effects" (= an average result in the phenotype considered over a large sample of gene-backgrounds and environments). These functional forms are "alleles" (or "allelomorphs", the original term). If both alleles at a gene have the same phenotypic effect (are the "same"), the gene is said to be "homozygous": if each allele at a gene is different in effect, the gene is "heterozygous".

The average phenotypic outcome may also depend upon how alleles interact with their own homologous partner in the disomic genotype ("dominance"), and on how these alleles interact with those of other genes at other loci which also affect this phenotypic trait ("epistasis"). Notice that we have combined classical genetics ideas with those of statistics in this exposition. Terms such as "gene", "homologue", "allele", "homozygous" and "heterozygous" are genetical: but "effect" is statistical, and refers to the average observed over an infinity of backgrounds, both genetical and environmental.

Thus, we have very sneakily defined the "genetic value" (or "genotypic" value) as the infinity mean of all the *phenotypes* it can ever produce in time and space! Before molecular genetics, there simply was no other way to do it! And after molecular genetics, this is still the most utilitarian way to tackle the idea of genetic value! Also notice that we have openly used fundamentals of reproductive biology behind the genetics.

The founder of Quantitative Genetics - Sir Ronald Fisher - perceived all of this when he proposed the first mathematics of this branch of Genetics [Fisher R.A. (1930). The Genetical theory of Natural Selection. Oxford Clarendon Press.]. He sought to define a single statistical summary of all the variance arising from phenotypic change during the course of genetical assortment and segregation, which he called the "genetic" variance. His residual genotypic variance (which he called simply the "residual") represented that part of assortment which did not lead to phenotypic change, although the genes themselves had in fact been subject to meiosis and syngamy, of course.

[A simplified exposition of this can be found in Falconer and Mackay (1966) - see References.] These partitions subsequently became the familiar subdivisions of the "additive" (A) and "dominance" (D) variances, respectively. These later names are utterly misleading and very unfortunate, and have led to much confusion as to what they mean genetically. [It would have been better for posterity had they been named "Assortative" (A), and "Stable" (S).] A more gene-focused partitioning was invented by Mather and Jinks in 1971, but they also were statisticians rather than geneticists, had their own rather opqaque symbolism, and became somewhat overwhelmed by the Fisherian approach. Added to all of this was the problem that Fisher's underlying reproduction model (mating system) was unrealistically simplistic: whilst it facilitated solving the equations, it didn't describe many real-life scenarios.

Fortunately, Wright (in the 1950s) did provide the means to overcome complex mating systems, but its own complexity minimised its popular adoption. Early in this millennium [Heredity (2003) 91: 85-89], a comprehensive linking together of all of these approaches reconciled their various meanings and relationships (as well as correcting an error), and suggested that new partitions reflecting real homozygosis [a] and heterozygosis [d] should replace the present "additive" [A] and "dominance" [D] subdivisions. This paper revealed, by the way, that the Additive Genetical variance consists of all the homozygote variance, plus part of the dominance variance, and a frequency-weighted covariance between homozygote and heterozygote gene effects. The so-called Dominance variance contains only the remainder of the overall dominance variance of the gene in question, being therefore very misleading indeed! It should be understood, however, that either method of partitioning still accounts for all of the genotypic variance in the model being used: it's the way it has been divided up which is being debated. At least the Environmental variance is much more straight-forward. This can be subdivided into a pure environmental component (E) and an interaction component (I) describing the gene-environment interaction. The overall "single gene" model can be written as:

$$P = a + d + E + I.$$

Expansion of the model to multiple genes (loci) is still not resolved satisfactorily, and until that is solved it is not possible to account for epistasis. The problem is being tackled currently. The contribution of those components cannot be determined in a single individual, but they can be estimated for whole populations by estimating the variances for those components, denoted as:

$$V_P = V_a + V_d + V_E + V_I$$

The heritability of a trait is the proportion of the total (i.e. phenotypic) variance (V_P) that is explained by the total genotypic variance (V_G). This is known also as the "broad sense" heritability (H^2). If only Additive genetic variance (V_A) is used in the numerator, the heritability is "narrow sense" (h^2). Unfortunately, this is often simply called "heritability", with little reflection about its true meaning. The broadsense heritability indicates the genotypic determination of the phenotype: while the latter estimates the degree of assortative disequlibrium in the trait. Fisher proposed that this narrow-sense heritability might be appropriate in considering the results of natural selection, focusing as it does on disequilibrium: and it has been used also for predicting the results of artificial selection. This latter usage seems to be inappropriate, however, as breeders are interested in steering attributes towards new phenotypes (that is in utilising all the gene effects), rather than simply exploiting disequilibrium. But old dogmas die hard!

Resemblance between Relatives

Central in estimating the variances for the various components is the principle of relatedness. A child has a father and a mother. Consequently, the child and father share 50% of their alleles, as do the child and the mother. However, the mother and father normally do not share alleles as a result of shared ancestors. Similarly, two full siblings share also on average 50% of the alleles with each other, while half siblings share only 25% of their alleles. This variation in relatedness can be used to estimate which proportion of the total phenotypic variance (V_p) is explained by the above-mentioned components.

Correlated Traits

Although some genes have only an effect on a single trait, many genes have an effect on various traits. Because of this, a change in a single gene will have an effect on all those traits. This is calculated using covariances, and the phenotypic covariance (Cov_p) between two traits can be partitioned in the same way as the variances described above.

Genomics

Genomics is a discipline in genetics concerning the study of the genomes of organisms. The field includes intensive efforts to determine the entire DNA sequence of organisms and fine-scale genetic mapping efforts. The field also includes studies of intragenomic phenomena such as heterosis, epistasis, pleiotropy and other interactions between loci and alleles within the genome.

In contrast, the investigation of the roles and functions of single genes is a primary focus of molecular biology or genetics and is a common topic of modern medical and biological research. Research of single genes does not fall into the definition of genomics unless the aim of this genetic, pathway, and functional information analysis is to elucidate its effect on, place in, and response to the entire genome's networks.

For the United States Environmental Protection Agency, "the term "genomics" encompasses a broader scope of scientific inquiry associated technologies than when genomics was initially considered. A genome is the sum total of all an individual organism's genes. Thus, genomics is the study of all the genes of a cell, or tissue, at the DNA (genotype), mRNA (transcriptome), or protein (proteome) levels."

History

The first genomes to be sequenced were those of a virus and a mitochondrion, and were done by Fred Sanger. His group established techniques of sequencing, genome mapping, data storage, and bioinformatic analyses in the 1970-1980s. A major branch of genomics is still concerned with sequencing the genomes of various organisms, but the knowledge of full genomes has created the possibility for the field of functional genomics, mainly concerned with patterns of gene expression during various conditions. The most important tools here are microarrays and bioinformatics.

Study of the full set of proteins in a cell type or tissue, and the changes during various conditions, is called proteomics. A related concept is materiomics, which is defined as the study of the material properties of biological materials (e.g. hierarchical protein structures and materials, mineralized biological tissues, etc.) and their effect on the macroscopic function and failure in their biological context, linking processes, structure and properties at multiple scales through a materials science approach. The actual term 'genomics' is thought to have been coined by Dr. Tom Roderick, a geneticist at the Jackson Laboratory (Bar Harbor, ME) over beer at a meeting held in Maryland on the mapping of the human genome in 1986.

In 1972, Walter Fiers and his team at the Laboratory of Molecular Biology of the University of Ghent (Ghent, Belgium) were the first to determine the sequence of a gene: the gene for Bacteriophage MS2 coat protein. In 1976, the team determined the complete nucleotide-sequence of bacteriophage MS2-RNA. The first DNA-based genome to be sequenced in its entirety was that of bacteriophage Ö-174; (5,368 bp), sequenced by Frederick Sanger in 1977.

The first free-living organism to be sequenced was that of *Haemophilus influenzae* (1.8 Mb) in 1995, and since then genomes are being sequenced at a rapid pace.

As of September 2007, the complete sequence was known of about 1879 viruses, 577 bacterial species and roughly 23 eukaryote organisms, of which about half are fungi. Most of the bacteria whose genomes have been completely sequenced are problematic disease-causing agents, such as *Haemophilus influenzae*. Of the other sequenced species, most were chosen because they were well-studied model organisms or promised to become good models. Yeast (*Saccharomyces cerevisiae*) has long been an important model organism for the eukaryotic cell, while the fruit fly *Drosophila melanogaster* has been a very important tool (notably in early pre-molecular genetics).

The worm *Caenorhabditis elegans* is an often used simple model for multicellular organisms. The zebrafish *Brachydanio rerio* is used for many developmental studies on the molecular level and the flower

Arabidopsis thaliana is a model organism for flowering plants. The Japanese pufferfish (*Takifugu rubripes*) and the spotted green pufferfish (*Tetraodon nigroviridis*) are interesting because of their small and compact genomes, containing very little non-coding DNA compared to most species. The mammals dog (*Canis familiaris*), brown rat (*Rattus norvegicus*), mouse (*Mus musculus*), and chimpanzee (*Pan troglodytes*) are all important model animals in medical research.

Human Genomics

A rough draft of the human genome was completed by the Human Genome Project in early 2001, creating much fanfare. By 2007 the human sequence was declared "finished" (less than one error in 20,000 bases and all chromosomes assembled). Display of the results of the project required significant bioinformatics resources. The sequence of the human reference assembly can be explored using the UCSC Genome Browser.

Bacteriophage Genomics

Bacteriophages have played and continue to play a key role in bacterial genetics and molecular biology. Historically, they were used to define gene structure and gene regulation. Also the first genome to be sequenced was a bacteriophage. However, bacteriophage research did not lead the genomics revolution, which is clearly dominated by bacterial genomics. Only very recently has the study of bacteriophage genomes become prominent, thereby enabling researchers to understand the mechanisms underlying phage evolution. Bacteriophage genome sequences can be obtained through direct sequencing of isolated bacteriophages, but can also be derived as part of microbial genomes. Analysis of bacterial genomes has shown that a substantial amount of microbial DNA consists of prophage sequences and prophage-like elements. A detailed database mining of these sequences offers insights into the role of prophages in shaping the bacterial genome.

Cyanobacteria Genomics

At present there are 24 cyanobacteria for which a total genome sequence is available. 15 of these cyanobacteria come from the marine environment. These are six *Prochlorococcus* strains, seven marine *Synechococcus* strains, *Trichodesmium erythraeum* IMS101 and *Crocosphaera watsonii* WH8501. Several studies have demonstrated how these sequences could be used very successfully to infer important ecological and physiological characteristics of marine cyanobacteria.

However, there are many more genome projects currently in progress, amongst those there are further *Prochlorococcus* and marine *Synechococcus* isolates, *Acaryochloris* and *Prochloron*, the N_2-fixing filamentous cyanobacteria *Nodularia spumigena, Lyngbya aestuarii* and *Lyngbya majuscula*, as well as bacteriophages infecting marine cyanobaceria. Thus, the growing body of genome information can also be tapped in a more general way to address global problems by applying a comparative approach. Some new and exciting examples of progress in this field are the identification of genes for regulatory RNAs, insights into the evolutionary origin of photosynthesis, or estimation of the contribution of horizontal gene transfer to the genomes that have been analyzed.

Medical Genetics

Medical genetics is the specialty of medicine that involves the diagnosis and management of hereditary disorders. Medical genetics differs from Human genetics in that human genetics is a field of scientific research that may or may not apply to medicine, but medical geneics refers to the application of genetics to medical care. For example, research on the causes and inheritance of genetic disorders would be considered within both human genetics and medical genetics, while the diagnosis, management, and counseling of individuals with genetic disorders would be considered part of medical genetics.

In contrast, the study of typically non-medical phenotypes such as the gnetics of eye colour wuld be considered part of human genetics, but not necessarily relevant to medical genetics (except in situations such as albinism). *Genetic medicine* is a newer term for medical genetics and incorporates areas such as gene therapy, personalized medicine, and the rapidly emerging new medical specialty, predictive medicine.

Scope

Medical genetics encompasses many different areas, including clinical practice of physicians, genetic counselors, and nutritionists, clinical diagnostic laboratory activities, and research into the causes and inheritance of genetic disorders. Examples of conditions that fall within the scope of medical genetics include birth defects and dysmorphology, mental retardation, autism, metabolic and mitochondrial disorders, skeletal dysplasia, connective tissue disorders, cancer genetics, teratogens, and prenatal diagnosis. Medical genetics is increasingly becoming relevant to many common diseases. Overlaps with other medical speciaties are beginning to emerge, as recent advances in genetics are revealing etiologies for neurologic, endocrine, cardiovascular, pulmonary, ophthalmologic, renal, psychiatric, and dermatologic conditions.

Subspecialties

In some ways, many of the individual fields within medical genetics are hybrids between clinical care and research. This is due in part to recent advances in science and technology (for example, see the Human genome project) that have enabled an unprecedented understanding of genetic disorders.

Clinical Genetics: Clinical genetics is the practice of clinical medicine with particular attention to hereditary disorders. Referrals are made to genetics clinics for a variety of reasons, including birth defects, developmental delay, autism, epilepsy, short stature, and many others. Examples of genetic syndromes that are commonly seen in the genetics clinic include chromosomal rearrangements, Down syndrome, DiGeorge syndrome (22q11.2 Deletion Syndrome), Fragile X syndrome, Marfan syndrome, Neurofibromatosis, Turner syndrome, and Williams syndrome.

Metabolic/Biochemical Genetics: Metabolic (or biochemical) genetics involves the diagnosis and management of inborn errors of metabolism in which patients have enzymatic dficiencies that perturb biochemical pathways involved in metabolism of carbohydrates, amino acids, and lipids. Examples of metabolic disorders include galactosemia, glycogen storage disease, lysosomal storage disorders, metabolic acidosis, peroxisomal disorders, phenylketonuria, and urea cycle disorders.

Cytogenetics: Cytogenetics is the study of chromosomes and chromosome abnormalities. While cytogenetics historically relied on microscopy to analyze chromosomes, new molecular technologies such as array comparative genomic hybridization are now becoming widely used. Examples of chromosome abnormalities include aneuploidy, chromosomal rearrangements, and genomic deletion/duplication disorders.

Molecular Genetics: Molecular genetics involves the discovery of and laboratory testing for DNA mutations that underlie many single gene disorders. Examples of single gene disorders include achondroplasia, cystic fibrosis, Duchenne muscular dystrophy, hereditary breast cancer (BRCA1/2), Huntington disease, Marfan syndrome, Noonan syndrome, and Rett syndrome. Molecular tests are also used in the diagnosis of syndromes involving epigenetic abnormalities, such as Angelman syndrome, Beckwith-Wiedemann syndrome, Prader-willi syndrome, and uniparental disomy.

Mitochondrial Genetics: Mitochondrial genetics concerns the diagnosis and management of mitochondrial disorders, which have a molecular basis but often result in biochemical abnormalities due to deficient energy production.

There exists some overlap between medical genetic diagnostic laboratories and molecular pathology.

Genetic Counseling

Genetic counseling is the process of providing information about genetic conditions, diagnostic testing, and risks in other family members, within the framework of nondirective counseling. Genetic counselors are non-physician members of the medical genetics team who specialize in fmily risk assessment and counseling of patients regarding genetic disorders.

The precise role of the genetic counselor varies somewhat depending on the disorder.

History

Although genetics has its roots back in the 19th century with the work of the Bohemian monk Gregor Mendel and other pioneering scientists, huan genetics emerged later. It started to develop, albeit slowly, during the first half of the 20th century.

Mendelian (single-gene) inheritance was studied in a number of important disorders such as albinism, brachydactyly (short fingers and toes), and hemophilia. Mathematical approaches were also devised and applied to human genetics. Population genetics was created.

Medical genetics was a late developer, emerging largely after the close of World War II (1945) when the eugenics movement had fallen into disrepute. The Nazi misuse of eugenics sounded its death knell. Shorn of eugenics, a scientific approach could be used and was applied to human and medical genetics. Medical genetics saw an increasingly rapid rise in the second half of the 20th century and continues in the 21st century.

Notable Practitioners

- Victor McKusick (1921–2008), "The Father of Genetic Medicine" (http://www.ashg.org/pages/mckusick.shtml)

Current Practice

The clinical setting in which patients are evaluated determines the scope of practice, diagnostic, and therapeutic interventions. For the purposes of general discussion, the typical encounters between patients and genetic practitioners may involve:

- Referral to an out-patient genetics clinic (pediatric, adult, or combined) or an in-hospital consultation, most often for diagnostic evaluation.

- Specialty genetics clinics focusing on management of inborn errors of metabolism, skeletal dysplasia, or lysosomal storage diseases.

- Referral for counseling in a prenatal genetics clinic to discuss risks to the pregnancy (advanced maternal age, teratogen exposure, family history of a genetic disease), test results (abnormal maternal serum screen, abnormal ultrasound), and/ or options for prenatal diagnosis (typically amniocentesis or chorionic villus sampling).

- Multidisciplinary specialty clinics that include a clinical geneticist or genetic counselor (cancer genetics, cardiovascular genetics, craniofacial or cleft lip/palate hearing loss clinics, muscular dystrophy/neurodegenerative disorder clinics).

Diagnostic Evaluation

Each patient will undergo a diagnostic evaluation tailored to their own particular presenting signs and symptoms. The geneticist will establish a differential diagnosis and recommend appropriate testing. Increasingly, clnicians use SimulConsult, paired with the National Library of Medicine Gene Review articles, to narrow the list of hypotheses (known as the differential diagnosis) and identify the tests that are relevant for a particular patient. These tests might evaluate for chromosomal disorders, inborn errors of metabolism, or single gene disorders.

Chromosome Studies: Chromosome studies are used in the general genetics clinic to determine a cause for developmental delay/ mental retardation, birth defects, dysmorphic features, and/or autism.

Chromosome analysis is also performed in the prenatal setting to determine whether a fetus is affected with aneuploidy or other chromosome rearrangements. Finally, chromosome abnormalities are often detected in cancer samples. A large number of different methods have been developed for chromosome analysis:

- Chromosome analysis using a karyotype involves special stains that generate light and dark bands, allowing identification of each chromosome under a microscope.

- Fluorescence in situ hybridization (FISH) involves fluorescent labeling of probes that bind to specific DNA sequences, used for identifying aneuploidy, genomic deletions or duplications, characterizing chromosomal translocations and determining the origin of ring chromosomes.

- Chromosome painting is a technique that uses fluorescent probes specific for each chromosome to differentially label each chromosome. This technique is more often used in cancer cytogenetics, where complex chromosome rearrangements can occur.

- Array comparative genomic hybridization is a new molecular technique that involves hybridization of an individual DNA sample to a glass slide or microarray chip containing molecular probes (ranging from large ~200kb bacterial artificial chromosomes to small oligonucleotides) that represent unique regions of the genome. This method is particularly sensitive for detection of genomic gains or losses across the genome but does not detect balanced translocations or distinguish the location of duplicated genetic material (for example, a tandem duplication versus an insertional duplication).

Basic Metabolic Studies: Biochemical studies are performed to screen for imbalances of metabolites in the bodily fluid, usually the blood (plasma/serum) or urine, but also in cerebrospinal fluid (CSF). Specific tests of enzyme function (either in leukocytes, skin fibroblasts, liver, or muscle) are also employed under certain circumstances. In the US, the newborn screen incorporates biochemical tests to screen for treatable conditions such as galactosemia and phenylketonuria (PKU). Patients suspected to have a metabolic condition might undergo the following tests:

- Quantitative amino acid analysis is typically performed using the ninhydrin reaction, followed by liquid chromatography to measure the amount of amino acid in the sample (either urine, plasma/serum, or CSF). Measurement of amino acids in plasma or serum is used in the evaluation of disorders of amino acid metabolism such as urea cycle disorders, maple syrup urine disease, and PKU. Measurement of amino acids in urine can be useful in the diagnosis of cystinuria or renal Fanconi syndrome as can be seen in cystinosis.

- Urine organic acid analysis can be either performed using quantitative or qualitative methods, but in either case the test is used to detect the excretion of abnormal organic acids. These compounds are normally produced during bodily metabolism of amino acids and odd-chain fatty acids, but accumulate in patients with certain metabolic conditions.

- The acylcarnitine combination profile detects compounds such as organic acids and fatty acids conjugated to carnitine. The test is used for detection of disorders involving fatty acid metabolism, including MCAD.

- Pyruvate and lactate are byproducts of normal metabolism, particularly during anaerobic metabolism. These compounds normally accumulate during exercise or ischemia, but are also elevated in patients with disorders of pyruvate metabolism or mitochondrial disorders.

- Ammonia is an end product of amino acid metabolism and is converted in the liver to urea through a series of enzymatic reactions termed the urea cycle. Elevated ammonia can therefore be detected in patients with urea cycle disorders, as well as other conditions involving liver failure.

- Enzyme testing is performed for a wide range of metabolic disorders to confirm a diagnosis suspected based on screening tests.

Molecular Studies:

- DNA sequencing is used to directly analyze the genomic DNA sequence of a particular gene. In general, only the parts of the gene that code for the expressed protein (exons) and small amounts of the flanking untranslated regions and introns are analyzed. Therefore, although these tests are highly specific and sensitive, they do not routinely identify all of the mutations that could cause disease.

- DNA methylation analysis is used to diagnose certain genetic disorders that are caused by disruptions of epigenetic mechanisms such as genomic imprinting and uniparental disomy.

- Southern blotting is an early technique basic on detection of fragments of DNA separated by size through gel electrophoresis and detected using radiolabeled probes. This test was routinely used to detect deletions or duplications in conditions such as Duchenne muscular dystrophy but is being replaced by high-resolution array comparative genomic hybridization techniques. Southern blotting is still useful in the diagnosis of disorders caued by trinucleotide repeats.

- Short tandem repeats are unique markers that can be used to determine haplotypes and are used in identity testing for maternal cell contamination.

Treatments: Each cell of the body contains the hereditary information (DNA) wrapped up in structures called chromosmes. Since genetic syndromes are typically the result of alterations of the chromosomes or genes, there is no treatment currently available that can correct the genetic alterations in every cell of the body. Therefore, there is currently no "cure" for genetic disorders.

However, for many genetic syndromes there is treatment available to manage the symptoms. In some cases, particularly inborn errors of metabolism, the mechanism of disease is well understood and offers the potential for dietary and medical management to prevent or reduce the long-term complications. In other cases, infusion therapy is used to replace the missing enzyme. Current research is actively seeking to use gene therapy or other new medications to treat specific genetic disorders.

Management of Metabolic Disorders: In general, metabolic disorders arise from enzyme deficiencies that disrupt normal metbolic pathways. Compound "A" is metabolized to "B" by enzyme "X", compound "B" is metabolized to "C" by enzyme "Y", and compound "C" is metabolized to "D" by enzyme "Z". If enzyme "Z" is missing, compound "D" will be missing, while compounds "A", "B", and "C" will build up. The pathogenesis of this particular condition could result from lack of compound "D", if it is critical for some cellular function, or from toxicity due to excess "A", "B", and/or "C". Treatment of the metabolic disorder could be achieved through dietary supplementation of compound "D" and dietary restriction of compounds "A", "B", and/ or "C" or by treatment with a medication that promoted disposal of excess "A", "B", or "C". Another approach that can be taken is enzyme replacement therapy, in which a patient is given an infusion of the missing enzyme.

- *Diet:* Dietary restriction and supplementation are key measures taken in several well-known metabolic disorders, including galactosemia, phenylketonuria (PKU), maple syrup urine disease, organic acidurias and urea cycle disorders. Such restrictive diets can be difficult for the patient and family to maintain, and require close consultation with a nutritionist who has special experience in metabolic disorders. The composition of the diet will change depending on the caloric needs of the growing child and special attention is needed during a pregnancy if a woman is affected with one of these disorders.

- *Medication:* Medical approaches include enhancement of residual enzyme activity (in cases where the enzyme is made but is not functioning properly), inhibition of other enzymes in the biochemical pathway to prevent buildup of a toxic compound, or diversion of a toxic compound to another form that can be excreted. Examples include the use of high doses of pyridoxine (vitamin B6) in some patients with homocystinuria to boost the activity of the residual cystathione synthase enzyme, administration of biotin to restore activity of several enzymes affected by deficiency of biotinidase, treatment with NTBC in Tyrosinemia to inhibit the production of succinylacetone which causes liver toxicity, and the use of sodium benzoate to decrease ammonia build-up in urea cycle disorders.

- *Enzyme Replacement Therapy:* Certain lysosomal storage diseases are treated with infusions of a recombinant enzyme (produced in a laboratory), which can reduce the accumulation of the compounds in various tissues. Examples include Gaucher disease, Fabry disease, Mucopolysaccharidoses and Glycogen storage disease type II. Such treatments are limited by the ability of the enzyme to reach the affected areas (the blood brain barrier prevents enzyme from reaching the brain, for example), and can sometimes be associated with allergic reactions. The long-term clinical effectiveness of enzyme replacement therapies vary widely among different disorders.

Other Examples:

- Angiotensin receptor blockers in Marfan syndrome & Loeys-Dietz
- Bone marrow transplantation
- Gene therapy

Ethical, Legal and Social Implications

Genetic information provides a unique type of knowledge about an individual and his/her family, fundamentally different than a typically laboratory test that provides a "snapshot" of an individual's health status. The unique status of genetic information and inherited disease has a number of ramifications with regard to ethical, legal, and societal concerns.

Societies: The more empirical approach to human and medical genetics was formalized by the founding in 1948 of the American Society of Human Genetics. The Society first began annual meetings that year (1948) and its international counterpart, the International Congress of Human Genetics, has met every 5 years since its inception in 1956. The Society publishes the American Journal of Human Genetics on a monthly basis. Medical genetics is now recognized as a distinct medical specialty in the U.S. with its own approved board (the American Board of Medical Genetics) and clinical specialty college (the American College of Medical Genetics). The College holds an annual scientific meeting, publishes a monthly journal, Genetics in Medicine, and issues position papers and clinical practice guidelines on a variety of topics relevant to human genetics.

Research: The broad range of research in medical genetics reflects the overall scope of this field, including basic research on genetic inheritance and the human genome, mechanisms of genetic and metabolic disorders, translational research on new treatment modalities, and the impact of genetic testing

Basic Genetics Research: Basic research geneticists usually undertake research in universities, biotechnology firms and research institutes.

Allelic Architecture of Disease: Sometimes the link between a disease and an unusual gene variant is more subtle. The genetic architecture of common diseases is an important factor in determining the extent to which patterns of genetic variation influence group differences in health outcomes. According to the common disease/ common variant hypothesis, common variants present in the ancestral population before the dispersal of modern humans from Africa play an important role in human diseases. Genetic variants associated with Alzheimer disease, deep venous thrombosis, Crohn disease, and type 2 diabetes appear to adhere to this model. However, the generality of the model has not yet been established and, in some cases, is in doubt. Some diseases, such as many common cancers, appear not to be well described by the common disease/common variant model. Another possibility is that common diseases arise in part through the action of combinations of variants that are individually rare.

Most of the disease-associated alleles discovered to date have been rare, and rare variants are more likely than common variants to be differentially distributed among groups distinguished by ancestry. However, groups could harbor different, though perhaps overlapping, sets of rare variants, which would reduce contrasts between groups in the incidence of the disease. The number of variants contributing to a disease and the interactions among those variants also could influence the distribution of diseases among groups.

The difficulty that has been encountered in finding contributory alleles for complex diseases and in replicating positive associations suggests that many complex diseases involve numerous variants rather than a moderate number of alleles, and the influence of any given variant may depend in critical ways on the genetic and environmental background.

If many alleles are required to increase susceptibility to a disease, the odds are low that the necessary combination of alleles would become concentrated in a particular group purely through drift.

Population Substructure in Genetics Research: One area in which population categories can be important considerations in genetics research is in controlling for confounding between population substructure, environmental exposures, and health outcomes. Association studies can produce spurious results if cases and controls have differing allele frequencies for genes that are not related to the disease being studied, although the magnitude of this problem in genetic association studies is subject to debate. Various methods have been developed to detect and account for population substructure, but these methods can be difficult to apply in practice.

Population substructure also can be used to advantage in genetic association studies. For example, populations that represent recent mixtures of geographically separated ancestral groups can exhibit longer-range linkage disequilibrium between susceptibility alleles and genetic markers than is the case for other populations. Genetic studies can use this admixture linkage disequilibrium to search for disease alleles with fewer markers than would be needed otherwise. Association studies also can take advantage of the contrasting experiences of racial or ethnic groups, including migrant groups, to search for interactions between particular alleles and environmental factors that might influence health.

Behavioural Genetics

Behavioural genetics is the field of study that examines the role of genetics in animal (including human) behaviour. Often associated with the "nature versus nurture" debate, behavioural genetics is highly interdisciplinary, involving contributions from biology, genetics, ethology, psychology, and statistics. Behavioural geneticists study the inheritance of behavioural traits. In humans this often use the twin study or adoption study. In animal studies, breeding, transgenesis, and gene knockout techniques are common; psychiatric genetics is a closely related field.

History

Sir Francis Galton, a nineteenth-century intellectual, is recognized as one of the first behavioural geneticists. Galton, a cousin of Charles Darwin, studied the heritability of human ability, focusing on mental characteristics as well as eminence among close relatives in the English upper-class. In 1869, Galton published his results in *Hereditary Genius*. In his work, Galton "introduced multivariate analysis and paved the way towards modern Bayesian statistics" that are used throughout the sciences—launching what has been dubbed the "Statistical Enlightenment".

Behaviour genetics, *per-se*, gained recognition as a research discipline with the publication in 1960 of the textbook *Behaviour Genetics* by J.L. Fuller and W.R. Thompson.

Underscoring the role of evolution in behavioural genetics, Theodosius Dobzhansky was elected the first president of the Behaviour Genetics Association in 1972; the BGA bestows the Dobzhansky Award on researchers for their outstanding contributions to the field. In the early 1970s, Lee Ehrman, a doctoral student of Dobzhansky, wrote seminal papers describing the relationship between genotype frequency and mating success in Drosophila, lending impetus to the pursuit of genetic studies of behaviour in other animals.

Notable Behavioural Geneticists

Notable behavioural geneticists include Dorret Boomsma, John DeFries, Lindon Eaves, David Fulker, John Hewitt, Kenneth Kendler, John Loehlin, Nick Martin, Gerald McClearn, Robert Plomin, Theodore Reich, who was a pioneer in psychiatric genetics, Hans van Abeelen, Avshalom Caspi, and Steven G. Vandenberg, the founding editor of the journal *Behaviour Genetics*.

Genetic Diversity

Genetic diversity, the level of biodiversity, refers to the total number of genetic characteristics in the genetic makeup of a species. It is distinguished from genetic variability, which describes the tendency of genetic characteristics to vary. Genetic diversity serves as a way for populations to adapt to changing environments. With more variation, it is more likely that some individuals in a population will possess variations of alleles that are suited for the environment. Those individuals are more likely to survive to produce offspring bearing that allele. The population will continue for more generations because of the success of these individuals. The academic field of population genetics includes several hypotheses and theories regarding genetic diversity. The neutral theory of evolution proposes that diversity is the result of the accumulation of neutral substitutions. Diversifying selection is the hypothesis that two subpopulations of a species live in different environments that select for different alleles at a particular locus. This may occur, for instance, if a species has a large range relative to the mobility of individuals within it. Frequency-dependent selection is the hypothesis that as alleles become more common, they become more vulnerable. This is often invoked in host-pathogen interactions, where a high frequency of a defensive allele among the host means that it is more likely that a pathogen will spread if it is able to overcome that allele.

Importance of Genetic Diversity

There are many different ways to measure genetic diversity. The modern causes for the loss of animal genetic diversity have also been studied and identified. A 2007 study conducted by the National Science

Foundation found that genetic diversity and biodiversity are dependent upon each other—that diversity within a species is necessary to maintain diversity among species, and vice versa. According to the lead researcher in the study, Dr. Richard Lankau, "If any one type is removed from the system, the cycle can break down, and the community becomes dominated by a single species."

The interdependence between genetic and biological diversity is delicate. Changes in biological diversity lead to changes in the environment, leading to adaptation of the remaining species. Changes in genetic diversity, such as in loss of species, leads to a loss of biological diversity.

Survival and Adaptation

Genetic diversity plays a very important role in survival and adaptability of a species because when a species's environment changes, slight gene variations are necessary to produce changes in the organisms' anatomy that enables it to adapt and survive. A species that has a large degree of genetic diversity among its population will have more variations from which to choose the most fit alleles. Increase in genetic diversity is also essential for a species to evolve. Species that have very little genetic variation are at a great risk. With very little gene variation within the species, healthy reproduction becomes increasingly difficult, and offspring often deal with similar problems to those of inbreeding. The vulnerability of a population to certain types of diseases can also increase with reduction in genetic diversity.

Agricultural Relevance

When humans initially started farming, they used selective breeding to pass on desirable traits of the crops while omitting the undesirable ones. Selective breeding leads to monocultures: entire farms of nearly genetically identical plants. Little to no genetic diversity makes crops extremely susceptible to widespread disease. Bacteria morph and change constantly. When a disease causing bacterium changes to attack a specific genetic variation, it can easily wipe out vast quantities of the species. If the genetic variation that the bacterium is best at attacking happens to be that which humans have selectively bred to use for harvest, the entire crop will be wiped out.

A very similar occurrence is the cause of the infamous Potato Famine in Ireland. Since new potato plants do not come as a esult of reproductin but rather frompieces of the parent lant, no genetic diversty is developed, and the entire crop is essentially a clone of one potato, it is especially susceptible to an epidemic. In the 1840s, much of Ireland's population depended on potatoes for food. They planted namely the "lumper" variety of potato, which was susceptible to a rot-causing plasmodiophorid called *Phytophthora infestans*. This plasmodiophorid destroyed the vast majority of the potato crop, and left one million people to starve to death.

Coping with Poor Genetic Diversity

The natural world has several ways of preserving or increasing genetic diversity. Among oceanic plankton, viruses aid in the genetic shifting process. Ocean viruses, which infect the plankton, carry genes of other organisms in addition to their own. When a virus containing the genes of one cell infects another, the genetic makeup of the latter changes. This constant shift of genetic make-up helps to maintain a healthy population of plankton despite complex and unpredictable environmental changes.

Cheetahs are a threatened species. Extremely low genetic diversity and resulting poor sperm quality has made breeding and survivorship difficult for cheetahs –- only about 5% of cheetahs survive to adulthood. About 10,000 years ago, all but the *jubatus* species of cheetahs died out. The species encountered a population bottleneck and close family relatives were forced to mate with each other, or inbreed. However, it has been recently discovered that female cheetahs can mate with more than one male per litter of cubs. They undergo induced ovulation, which means that a new egg is produced every time a female mates. By mating with multiple males, the mother increases the genetic diversity within a single litter of cubs.

Measures of Genetic Diversity

Genetic Diversity of a population can be assessed by some simple measures.

- Gene Diversity is the proportion of polymorphic loci across the genome.
- Heterozygosity is the mean number of individuals with polymorphic loci.
- Alleles per locus is also used to demonstrate variability.

Other Measures of Diversity

Alternatively, other types of diversity may be assessed for organisms:

- taxonomic diversity

- ecological diversity

- morphological diversity

There are broad correlations between different types of diversity. For example, there is a close link between vertebrate taxonomic and ecological diversity.

Genetic Erosion

Genetic erosion is a process whereby an already limited gene pool of an endangered species of plant or animal diminishes even more when individuals from the surviving population die off without getting a chance to meet and breed with others in their endangered low population. Genetic erosion occurs because each individual organism has many unique genes which get lost when it dies without getting a chance to breed. Low genetic diversity in a population of wild animals and plants leads to a further diminishing gene pool, inbreeding and a weakening immune system and fast tracks that species towards eventual extinction. All the world's endangered species are plagued by varying degrees of genetic erosion and most need a human assisted breeding programme to keep their population viable and to keep them from going extinct in the long run.

The more critically endangered the species is (the smaller the popuation is), the more magnified the effect of genetic erosion gets when each surviving individual of the species is lost without getting a chance to breed. Genetic erosion gets compounded and accelerated by habitat fragmentation, today most endangered species live in smaller and smaller chunks of fragmented habitat interspersed with human settlements and farmland making it impossible for them to naturally meet and breed with others of their kind, many die off without getting a chance to breed and pass on their genes in the living population. The gene pool of a species or a population is the complete set of unique alleles that would be found by inspecting the genetic material of every living member of that species or population.

A large gene pool indicates extensive genetic diversity, which is associated with robust populations that can survive bouts of intense selection. Meanwhile, low genetic diversity can cause reduced biological fitness and an increased chance of extinction.

Processes and Consequences

"A population bottleneck creates a shrinking gene pool that leaves fewer and fewer mating partners. What are the genetic implications?

The animals become part of a high stakes poker game — with a crooked dealer. After beginning with a 52-card deck, the players wind up with, say, five cards that they are dealt over and over. As they begin to inbreed, congenital effects appear, both physical and reproductive. Often abnormal sperm increase; infertility rises; the birthrate falls. Most perilous in the long run, each animal's immune defense system is weakened. Thus, even if an endangered species in a bottleneck can withstand whatever human development may be eating away at its habitat, it still faces the threat of an epidemic that could well be fatal to the entire population."

Genetic Erosion in Agricultural and Livestock Biodiversity

Genetic erosion in agricultural and livestock biodiversity is the loss of genetic diversity, including the loss of individual genes, and the loss of particular combinants of genes (or gene complexes) such as those manifested in locally adapted landraces of domesticated animals or plants adapted to the natural environment in which they originated. The term genetic erosion is sometimes used in a narrow sense, such as for the loss of alleles or genes, as well as more broadly, referring to the loss of varieties or even species. The major driving forces behind genetic erosion in crops are: variety replacement, land clearing, overexploitation of species, population pressure, environmental degradation, overgrazing, policy and changing agricultural systems.

The main factor, however, is the replacement of local varieties of domestic plants and animals by high yielding or exotic varieties or species. A large number of varieties can also often be dramatically reduced when commercial varieties (including GMOs) are introduced into traditional farming systems. Many researchers believe that the main problem related to agro-ecosystem management is the general tendency towards genetic and ecological uniformity imposed by the development of modern agriculture.

Human Intervention, Modern Science and Safeguards to Guard Against Genetic Erosion

In-Situ Conservation: With advances in modern science several tecniques and safeguards have emerged to check the relentless advance of genetic erosion and the resulting acceleration of endangered species towards extinction. However many of these techniques and safeguards are too expensive yet to be practical, the best way to protect species is to protect their habitat and to let them live in it naturally.

Wildlife sanctuaries and national parks have been created to preserve entire ecosystems with all the web of species which call them home. Wildlife corridors are created to join fragmented habitats to enable endangered species to travel, meet and breed with others of their kind. Scientific conservation and modern wildlife management techniques with the help of scientifically trained staff help manage these protected ecosystems and the wildlife found in them. Wild animals are also translocated and reintroduced to other locations physically when fragmented wildlife habitat is too far and isolated to be able to link it with a wildlife corridor or when local extinction has already occurred.

Ex-Situ Conservation

Modern policies of the zoo associations and zoos around the world have changed to putting extreme importance on keeping and breeding wild sourced pure species and subspecies of animals and birds in their registered endangered species breeding programmes which will have a chance to be reintroduced and survive in the wild. Main objectives of zoos today has changed to breed pure breed species and subspecies to assist conservation efforts in the wild. Zoos do this by maintaining extremely detailed scientific breeding records i.e. studbooks and loaning their pure breed wild animals and birds to other zoos around the country and indeed globally for breeding to safeguard against inbreeding and hybrids which are considered genetically compromised thus not fit for reintroduction in the wild and in the case of unnaturally found hybrids also to guard against genetic pollution in naturally evolved, region specific, pure wild stocks.

Costly and sometimes controversial ultra modern ex-situ conservation techniques have emerged for saving the genetic biodiversity on our planet and the diversity in their gene pool by guarding against genetic erosion through modern concepts like seedbanks, sperm banks and tissue banks. Sperms, eggs and embryos can now be frozen and kept in these banks which are sometimes called Modern Noah's Ark or Frozen Zoos. Cryopreservation techniques are used to freeze these living materials and keep them alive by storing them submerged in liquid nitrogen tanks. Thus, preserved material can then be used for artificial insemination, in vitro fertilization, embryo transfer and cloning to protect diversity in the gene pool of critically endangered species.

It is today possible to save endangered species from extinction by preserving parts like tissue, sperms, eggs etc. even after the death of a critically endangered animal or collected from one found freshly dead in captivity or from wild and resurrect it with the help of cloning and give it another chance to breed its genes into the living population of the respective species which is threatened with extinction. Resurrection of dead critically endangered wildlife with the help of cloning is still being perfected and is still too expensive to be practical but with time and advancement is science it may well become a routine procedure.

Recently strategies have been made to find an integrated approach to in situ and ex situ conservation.

Genetic Hitchhiking

Genetic hitchhiking is the process by which an evolutionarily neutral or deleterious allele or mutation may spread through the gene pool by virtue of being linked to a gene that is positively selected. More generally, genetic hitchhiking can refer to the process by which a gene's frequency changes due to selection operating upon linked genes. Proximity on a chromosome may allow genes to be dragged through the selection process due to an advantageous gene nearby. The most often cited example of this is one of a hypothetical mutator whih increases the general mutation rate in the area around it.

—M———A—

On this chromosome the gene M is a mutator allele, increasing the rate of mutation in the surrounding area. A is an allele which is at fixation in the population - that is, all individuals possess this allele in the homozygous state. Due to the increased mutation rate, the A allele may be mutated into a new, advantageous allele, A*.

—M———A*—

The individual in which this chromosome lies will now have a selective advantage over other individuals of this species, so the allele A* will spread through the population by the normal processes of natural selection. M, due to its proximity to A*, will be dragged through into the general population. This process only works when M is very close to the allele it has mutated. A greater distance would increase the chance of recombination separating M from A*, leaving M alone with any deleterious mutations it may have caused. For this reason the two alleles are generally directly next to each other, or present in asexual species where recombination cannot disrupt linkage.

In the example above, the mutator caused the adaptive allele to appear, but genetic hitchhiking can also happen by chance alone.

Whether a neutral allele becomes fixed is a matter of chance. The traditional view of this stochastic process is that it is dominated by sampling error, that is genetic drift. But it may instead be dominated by whether the mutation is linked to a good genetic background: this is known as genetic draft.

Genetic Monitoring

Genetic monitoring is the use of molecular markers to (i) identify individuals, species or populations, or (ii) to quantify changes in population genetic metrics (such as effective population size, genetic diversity and population size) over time. Genetic monitoring can thus be used to detect changes in species abundance and/or diversity, and has become an important tool in both conservation and livestock management. The types of molecular markers used to monitor populations are most commonly mitochondrial, microsatellites or single-nucleotide polymorphisms (SNPs), while earlier studies also used allozyme data. Species gene diversity is also recognized as an important biodiversity metric for implementation of the Convention on Biological Diversity.

Types

Types of population changes that can be detected by genetic monitoring include population growth and decline, spread of pathogens, adaptation to environmental change, hybridization, introgression and fragmentation events. Most of these changes are monitored using 'neutral' genetic markers (markers for which mutational changes do not change their adaptive fitness within a population). However markers showing adaptive responses to environmental change can be 'non-neutral' (e.g. mutational changes affect their relative fitness within a population).

Two broad categories of genetic monitoring have been defined: Category I encompasses the use of genetic markers as identifiers of individuals (Category Ia), populations and species (Category Ib) for traditional population monitoring. Category II represents the use of genetic markers to monitor changes of population genetic parameters, which include estimators of effective population size (Ne), genetic variation, population inter-mixing, structure and migration.

Examples

Estimating Abundance and Life History Parameters – Category Ia: At the individual level, genetic identification can enable estimation of population abundance and population increase rates within the framework of mark-recapture models. The abundance of cryptic or elusive species that are difficultto monitor can be estimated by collecting non-invasive biological samples in the field (e.g. feathers, scat or fur) and using these to identify individuals through microsatellite or single-nucleotide polymorphism (SNP) genotyping. This census of individuals can then be used to estimate population abundance via mark-recapture analysis. For example, this technique has been used to monitor populations of grizzly bear, Brush-tailed Rock-wallaby, Bengal tiger and snow leopard. Population growth rates are a product of rates of population recruitment and survival, and can be estimated through open mark-recapture models. For example, DNA from feathers shed by the Eastern Imperial Eagle shows lower cumulative survival over time than seen for other long-lived raptors.

Identifying Species – Category Ib: Use of molecular genetic techniques to identify species can be useful for a number of reasons. Species identification in the wild can be used to detect changes in population ranges or site occupancy, rates of hybridization and the emergence and spread of pathogens and invasive species. Changes in population ranges have been investigated for Iberian lynx and wolverine, while monitoring of Westslope cutthroat trout shows widespread ongoing hybridization with introduced rainbow trout and Canada lynx-bobcat hybrids have been detected at the southern periphery of the current population range for lynx. The emergence and spread of pathogens can be tracked using diagnostic molecular assays – for example, identifying the spread of West Nile virus among mosquitoes in the eastern US to identify likely geographical origins of infection and identifying gene loci associated with parasite susceptibility in bighorn sheep. Genetic monitoring of invasive species is of conservation and economic interest, as invasions often affect the ecology and range of native species and may also bring risks of hybridization (e.g. for copepods, barred owl and spotted owl, and Lessepsian rabbitfish).

Species identification is also of considerable utility in monitoring fisheries and wildlife trade, where conventional visual identification of butchered or flensed products is difficult or impossible. Monitoring of trade and consumption of species of conservation interest can be carried out using molecular amplification and identification of meat or fish obtained from markets. For example genetic market surveys have been used to identify protected species and populations of whale (e.g., North Pacific Minke whale) and dolphin species appearing in the marketplace. Other surveys of market trade have focused on pinnipeds, sea horses and sharks. Such surveys are used to provide ongoing monitoring of the quantity and movement of fisheries and wildlife products through markets and for detecting poaching or other illegal, unreported or unregulated (IUU) exploitation (e.g. IUU fishing).

Although initial applications focused on species identification and population assessments, market surveys also provide the opportunity for a range of molecular ecology investigations including capture-recapture, assignment tests and population modeling. These developments are potentially relevant to genetic monitoring Category II.

Monitoring Population Genetic Parameters – Category II: Monitoring of population changes through genetic means can be done retrospectively, through analysis of 'historical' DNA recovered from museum-archived species and comparison with contemporary DNA of that species. It can also be used as a tool for evaluating ongoing changes in the status and persistence of current populations. Genetic measures of relative population change include changes in diversity (e.g. heterozygosity and allelic richness). Monitoring of relative population changes through these metrics has been performed retrospectively for Beringian bison, Galapagos tortoise, houting, Atlantic salmon, northern pike, New Zealand snapper, steelhead trout, Greater Prairie Chicken, Mauritius kestrel and Hector's Dolphin and is the subject of many ongoing studies, including Danish and Swedish brown trout populations. Measuring absolute population changes (e.g. effective population size (Ne)) can be carried out by measuring changes in population allele frequencies ('Ftemporal') or levels of linkage disequilibrium over time ('LDNe'), while changing patterns of gene flow between populations can also be monitored by estimating differences in allele frequencies between populations over time. Subjects of such studies include grizzly bears, cod, red deer, Leopard frogs and Barrel Medic.

Genetic monitoring has also been increasingly used in studies that monitor environmental changes through changes in the frequency of adaptively selected markers. For example the genetically controlled photo-periodic response (hibernating time) of pitcher-plant mosquitos (*Wyeomyia smithii*) has shifted in response to longer growing seasons for pitcher plants brought on by warmer weather. Experimental wheat populations grown in contrasting environments over a period of 12 generations found that changes in flowering time were closely correlated with regulatory changes in one gene, suggesting a pathway for genetic adaptation to changing climate in plants. Genetic monitoring is also useful in monitoring the ongoing health of small, relocated populations. Good examples of this are found for New Zealand birds, many species of which were greatly impacted by habitat destruction and the appearance of numerous mammalian predators in the last century and have recently become part of relocation programmes that transfer a few 'founder' individuals to predator-free offshore "ecological" islands. E.g. Black robins, kakapo.

Status of Genetic Monitoring in Science

In February 2007 an international summit was held at the Institute of the Environment at UCLA, concerning 'Evolutionary Change in Human Altered Environments: An International Summit to translate Science into Policy'. This led to a special issue of the journal of Molecular Ecology organized around our understanding of genetic effects in three main categories: (i) habitat disturbance and climate change (ii) exploitation and captive breeding (iii) invasive species and pathogens.

In 2007 a Working Group on Genetic Monitoring was launched with joint support from NCEAS and NESCent to further develop the techniques involved and provide general monitoring guidance for policy makers and managers.

Genetic Monitoring in Natural Resource Agencies

Many natural resource agencies see genetic monitoring as a cost-effective and defensible way to monitor fish and wildlife populations. As such scientists in the U.S. Geologic Survey, U.S. Forest Service, National Parks Service, and National Marine Fisheries Service have been developing new methods and tools to use genetic monitoring, and applying such tools across broad geographic scales.

Genetic Pollution

Genetic pollution is undesirable and uncontrolled gene flow into wild populations.

Usage

- Conservation biologists and conservationists have long used genetic pollution as a term to describe gene flow from a domestic, feral, non-native or invasive species to a wild indigenous population.

- The term is of late being associated with the gene flow from a genetically engineered (GE) or genetically modified organism (GMO) to a non GE/GM organism.

Invasive Species

Conservation biologists and conservationists have from long before been using the term to describe the undesirable gene flow from domestic, feral, non-native and invasive species into wild indigenous species. For example, TRAFFIC is the international wildlife trade monitoring network which works to ensure that trade in wild plants and animals is not a threat to the conservation of nature.

They promote the awareness of the harmful effects of introduced invasive species that may *"hybridize with native species, causing genetic pollution"*. The Joint Nature Conservation Committee (JNCC) is the statutory adviser to the Government of United Kingdom and international nature conservation. Its work contributes to maintaining and enriching biological diversity and educating about the harmful effects of the introduction of invasive/non-native species. In this context they have advised that invasive species:

> "will alter the genetic pool (a process called genetic pollution), which is an irreversible change."

Effect upon Endangered Species

Compromise of certain wild plant or animal endangered species may have particular significance in the case of these relatively small populations of threatened taxa. As an example the endangered African wild dog has its conservation status undermined not only by loss of habitat, but also by admixture of gene flow from domesticated canids, as domesticated dogs interbreed with the wild species.

Genetic Engineering

In the field of agriculture, agroforestry and animal husbandry *genetic pollution* is being used to describe the undesirable gene flow between GE species and wild relatives;. An early use of the term *genetic pollution* in this later sense appears in a wide-ranging review of the potential ecological risks from genetic engineering in The Ecologist magazine in July 1989. It was also popularized by environmentalist Jeremy Rifkin in his 1998 book *The Biotech Century*. While intentional crossbreeding between two genetically distinct varieties is described as hybridization with the subsequent introgression

of genes, Rifkin, who had played a leading role in the ethical debate
for over a decade before, used genetic pollution to describe the risks
that might occur due the unintentional process of genetically modified
organisms (GMOs) dispersing their genes into the natural environment
by breeding with wild plants or animals.

The usage of genetic pollution by the Food and Agriculture
Organization of the United Nations (FAO) is currently defined as:

> *"Uncontrolled spread of genetic information (frequently*
> *referring to transgenes) into the genomes of organisms*
> *in which such genes are not present in nature."*

Since 2005 exist a GM Contamination Register, launched for
GeneWatch UK and Greenpeace International and, it record all
incidents intentional or accidental release of genetically modified
(GM) organisms.

In a 10 year study of four different crops, none of the genetically
modified plants were found to be more invasive or more persistent
than their conventional counterparts. An often cited example of genetic
pollution is the reputed discovery of transgenes from GE maize in
landraces of maize in Oaxaca, Mexico. The report from Quist and
Chapela, has since been discredited on methodological grounds. The
scientific journal that originally published the study concluded that
"the evidence available is not sufficient to justify the publication of
the original paper." More recent attempts to replicate the original
studies have concluded that genetically modified corn is absent from
southern Mexico in 2003 and 2004. The goal of genetic engineering
crop plants to help advance tenability and condition of the world food
supply has been a conlict with public and health concerns raised about
the safety of the food from the end product.

A 2004 study performed near an Oregon field trial for a genetically
modified variety of creeping bentgrass (*Agrostis stolonifera*) revealed
that the transgene and its associate trait (resistance to the glyphosate
herbicide) could be transmitted by wind pollination to resident plants
of different *Agrostis* species, up to 14 km from the test field. In 2007,
the Scotts Company, producer of the genetically modified bentgrass,
agreed to pay a civil penalty of $500,000 to the United States
Department of Agriculture (USDA). The USDA alleged that Scotts
"failed to conduct a 2003 Oregon field trial in a manner which ensured
that neither glyphosate-tolerant creeping bentgrass nor its offspring
would persist in the environment".

Controversial Term

Whether genetic pollution or similar terms, such as *"genetic deterioration"*, *"genetic swamping"*, *"genetic takeover"* and *"genetic aggression"*, are an appropriate scientific description of the biology of invasive species is debated Hymer and Simberloff argue that these types of terms:

> *"...imply either that hybrids are less fit than the parentals, which need not be the case, or that there is an inherent value in "pure" gene pools".*

They recommend that gene flow from invasive species be termed genetic mixing since:

> *" "Mixing" need not be value-laden, and we use it here to denote mixing of gene pools whether or not associated with a decline in fitness".*

Environmentalists such as Patrick Moore, an ex-member and cofounder of Greenpeace, questions if the term genetic pollution is more political than scientific. The term is considered to arouse emotional feelings towards the subject matter. In an interview he comments:

> *"If you take a term used quite frequently these days, the term "genetic pollution," otherwise referred to as genetic contamination, it is a propaganda term, nt a technical or scientific term. Pollution and contamination are both value judgments. By using the word "genetic" it gives the public the impression that they are talking about something scientific or technical—as if there were such a thing as genes that amount to pollution."*

Population Genetics

Population genetics is the study of allele frequency distribution and change under the influence of the four main evolutionary processes: natural selection, genetic drift, mutation and gene flow. It also takes into account the factors of recombination, population subdivision and population structure.

It attempts to explain such phenomena as adaptation and speciation. Population genetics was a vital ingredient in the emergence of the modern evolutionary synthesis. Its primary founders were Sewall Wright, J. B. S. Haldane and R. A. Fisher, who also laid the foundations for the related discipline of quantitative genetics.

Fundamentals

Population genetics is the study of the frequency and interaction of alleles and genes in populations. A population is a set of organisms in which any pair of members can breed together. This implies that all members belong to the same species and live near each other.

For example, all of the moths of the same species living in an isolated forest are a population. A gene in this population may have several alternate forms, which account for variations between the phenotypes of the organisms.

An example might be a gene for colouration in moths that has two alleles: black and white. A gene pool is the complete set of alleles for a gene in a single population; the allele frequency for an allele is the fraction of the genes in the pool that is composed of that allele (for example, what fraction of moth colouration genes are the black allele). Evolution occurs when there are changes in the frequencies of alleles within a population; for example, the allele for black colour in a population of moths becoming more common.

Hardy–Weinberg Principle

To understand the mechanisms that cause a population to evolve, it is useful to consider what conditions are required for a population not to evolve. The *Hardy-Weinberg principle* states that the frequencies of alleles (variations in a gene) in a sufficiently large population will remain constant if the only forces acting on that population are the random reshuffling of alleles during the formation of the sperm or egg, and random combination of the alleles in these sex cells during fertilization. Such a population is said to be in *Hardy-Weinberg equilibrium* as it is not evolving.

Hardy Weinberg equilibrium is impossible in nature. Genetic equilibrium is an ideal state that provides a baseline to measure genetic change against. Allele frequencies in a population remain static across generations, provided the following conditions are at hand: random mating, no mutation (the alleles don't change), no migration or emigration (no exchange of alleles between populations), infinitely large population size, and no selective pressure for or against any traits. In the simplest case of a single locus with two alleles: the dominant allele is denoted A and the recessve a and their frequenies are denoted y p and q; freq(A) = p; freq(a) = q; $p + q = 1$.

If the population is in equilibrium, then we will have freq(AA) = p^2 for the AA homozygotes in the population, freq(aa) = q^2 for the aa homozygotes, and freq(Aa) = $2pq$ for the heterozygotes. Based on these equations, useful but difficult-to-measure facts about a population can be determined. For example, a patient's child is a carrier of a recessive mutation that causes cystic fibrosis in homozygous recessive children. The parent wants to know the probability of her grandchildren inheriting the disease. In order to answer this question, the genetic counselor must know the chance that the child will reproduce with a carrier of the recessive mutation. This fact may not be known, but disease frequency is known. We know that the disease is caused by the homozygous recessive genotype; we can use the Hardy–Weinberg principle to work backward from disease occurrence to the frequency of heterozygous recessive individuals.

Scope and Theoretical Considerations

The mathematics of population genetics were originally developed as part of the modern evolutionary synthesis. According to Beatty (1986), it defines the core of the modern synthesis.

According to Lewontin (1974), the theoretical task for population genetics is a process in two spaces: a "genotypic space" and a "phenotypic space". The challenge of a *complete* theory of population genetics is to provide a set of laws that predictably map a population of genotypes (G_1) to a phenotype space (P_1), where selection takes place, and another set of laws that map the resulting population (P_2) back to genotype space (G_2) where Mendelian genetics can predict the next generation of genotypes, thus completing the cycle. Even leaving aside for the moment the non-Mendelian aspects of molecular genetics, this is clearly a gargantuan task.

In practice, there are two bodies of evolutionary theory that exist in parallel, traditional population genetics operating in the genotype space and the biometric theory used in plant and animal breeding, operating in phenotype space. The missing part is the mapping between the genotype and phenotype space. This leads to a "sleight of hand" (as Lewontin terms it) whereby variables in the equations of one domain, are considered parameters or *constants*, where, in a full-treatment they would be transformed themselves by the evolutionary process and are in reality *functions* of the state variables in the other domain. The "sleight of hand" is assuming that we know this mapping. Proceeding as if we do understand it is enough to analyze many cases of interest. For example, if the phenotype is almost one-to-one with genotype (sickle-cell disease) or the time-scale is sufficiently short, the "constants" can be treated as such; however, there are many situations where it is inaccurate.

The four Processes

Natural Selection: *Natural selection* is the process by which heritable traits that make it more likely for an organism to survive and successfully reproduce become more common in a population over successive generations.

The natural genetic variation within a population of organisms means that some individuals will survive more successfully than others in their current environment. Factors which affect reproductive success are also important, an issue which Charles Darwin developed in his ideas on sexual selection.

Natural selection acts on the phenotype, or the observable characteristics of an organism, but the genetic (heritable) basis of any phenotype which gives a reproductive advantage will become more common in a population. Over time, this process can result in adaptations that specialize organisms for particular ecological niches and may eventually result in the emergence of new species.

Natural selection is one of the cornerstones of modern biology. The term was introduced by Darwin in his groundbreaking 1859 book *On the Origin of Species,* in which natural selection was described by analogy to artificial selection, a process by which animals and plants with traits considered desirable by human breeders are systematically favoured for reproduction. The concept of natural selection was originally developed in the absence of a valid theory of heredity; at the time of Darwin's writing, nothing was known of modern genetics. The union of traditional Darwinian evolution with subsequent discoveries in classical and molecular genetics is termed the *modern evolutionary synthesis.* Natural selection remains the primary explanation for adaptive evolution.

Genetic Drift: *Genetic drift* is the change in the relative frequency in which a gene variant (allele) occurs in a population due to random sampling and chance. That is, the alleles in the offspring in the population are a random sample of those in the parents. And chance has a role in determining whether a given individual survives and reproduces. A population's allele frequency is the fraction or percentage of its gene copies compared to the total number of gene alleles that share a particular form.

Genetic drift is an important evolutionary process which leads to changes in allele frequencies over time. It may cause gene variants to disappear completely, and thereby reduce genetic variability. In contrast to natural selection, which makes gene variants more common or less common depending on their reproductive success, the changes due to genetic drift are not driven by environmental or adaptive pressures, and may be beneficial, neutral, or detrimental to reproductive success.

The effect of genetic drift is larger in small populations, and smaller in large populations. Vigorous debates wage among scientists over the relative importance of genetic drift compared with natural selection. Ronald Fisher held the view that genetic drift plays at the most a minor role in evolution, and this remained the dominant view for several decades. In 1968 Motoo Kimura rekindled the debate with his neutral theory of molecular evolution which claims that most of the changes in the genetic material are caused by genetic drift.

Mutation

Mutations are changes in the DNA sequence of a cell's genome and are caused by radiation, viruses, transposons and mutagenic chemicals, as well as errors that occr during meiosis or DNA replication. Errors are introduced particularly often in the process of DNA replication, in the polymerization of the second strand. These errors can also be induced by the organism itself, by cellular processes such as hypermutation.

Mutations can have an impact on the phenotype of an organism, especially if they occur within the protein coding sequence of a gene. Error rates are usually very low (1 error in every 10 million–100 million bases) due to the "proofreading" ability of DNA polymerases. Without proofreading, error rates are a thousandfold higher. Chemical damage to DNA occurs naturally as well, and cells use DNA repair mechanisms to repair mismatches and breaks in DNA. Nevertheless, the repair sometimes fails to return the DNA to its original sequence.

In organisms that use chromosomal crossover to exchange DNA and recombine genes, errors in alignment during meiosis can also cause mutations. Errors in crossover are especially likely when similar sequences cause partner chromosomes to adopt a mistaken alignment; this makes some regions in genomes more prone to mutating in this way. These errors create large structural changes in DNA sequence—

duplications, inversionsor deletions of entire regions, or the accidental exchanging of whole parts between different chromosomes (called translocation).

Mutation can result in several different types of change in DNA sequences; these can either have no effect, alter the product of a gene, or prevent the gene from functioning. Studies in the fly *Drosophila melanogaster* suggest that if a mutation changes a protein produced by a gene, this will probably be harmful, with about 70 percent of these mutations having damaging effects, and the remainder being either neutral or weakly beneficial.

Due to the damaging effects that mutations can have on cells, organisms have evolved mechanisms such as DNA repair to remove mutations. Therefore, the optimal mutation rate for a species is a trade-off between costs of a high mutation rate, such as deleterious mutations, and the metabolic costs of maintaining systems to reduce the mutation rate, such as DNA repair enzymes. Viruses that use RNA as their genetic material have rapid mutation rates, which can be an advantage since these viruses will evolve constantly and rapidly, and thus evade the defensive responses of e.g. the human immune system.

Mutations can involve large sections of DNA becoming duplicated, usually through genetic recombination. These duplications are a major source of raw material for evolving new genes, with tens to hundreds of genes duplicated in animal genomes every million years. Most genes belong to larger families of genes of shared ancestry. Novel genes are produced by several methods, commonly through the duplication and mutation of an ancestral gene, or by recombining parts of different genes to form new combinations with new functions.

Here, domains act as modules, each with a particular and independent function, that can be mixed together to produce genes encoding new proteins with novel properties. For example, the human eye uses four genes to make structures that sense light: three for colour vision and one for night vision; all four arose from a single ancestral gene.

Another advantage of duplicating a gene (or even an entire genome) is that this increases redundancy; this allows one gene in the pair to acquire a new function while the other copy performs the original function. Other types of mutation occasionally create new genes from previously noncoding DNA.

Gene Flow

Gene flow is the exchange of genes between populations, which are usually of the same species. Examples of gene flow within a species include the migration and then breeding of organisms, or the exchange of pollen. Gene transfer between species includes the formation of hybrid organisms and horizontal gene transfer.

Migration into or out of a population can change allele frequencies, as well as introducing genetic variation into a population. Immigration may add new genetic material to the established gene pool of a population. Conversely, emigration may remove genetic material. As barriers to reproduction between two diverging populations are required for the populations to become new species, gene flow may slow this process by spreading genetic differences between the populations. Gene flow is hindered by mountain ranges, oceans and deserts or even man-made structures such as the Great Wall of China, which has hindered the flow of plant genes.

Depending on how far two species have diverged since their most recent common ancestor, it may still be possible for them to produce offspring, as with horses and donkeys mating to produce mules. Such hybrids are generally infertile, due to the two different sets of chromosomes being unable to pair up during meiosis. In this case, closely related species may regularly interbreed, but hybrids will be selected against and the species will remain distinct. However, viable hybrids are occasionally formed and these new species can either have properties intermediate between their parent species, or possess a totally new phenotype. The importance of hybridization in creating new species of animals is unclear, although cases have been seen in many types of animals, with the gray tree frog being a particularly well-studied example.

Hybridization is, however, an important means of speciation in plants, since polyploidy (having more than two copies of each chromosome) is tolerated in plants more readily than in animals. Polyploidy is important in hybrids as it allows reproduction, with the two different sets of chromosomes each being able to pair with an identical partner during meiosis. Polyploids also have more genetic diversity, which allows them to avoid inbreeding depression in small populations.

Horizontal gene transfer is the transfer of genetic material from one organism to another organism that is not its offspring; this is most common among bacteria. In medicine, this contributes to the spread of antibiotic resistance, as when one bacteria acquires resistance genes it can rapidly transfer them to other species. Horizontal transfer of genes from bacteria to eukaryotes such as the yeast *Saccharomyces cerevisiae* and the adzuki bean beetle *Callosobruchus chinensis* may also have occurred.

An example of larger-scale transfers are the eukaryotic bdelloid rotifers, which appear to have received a range of genes from bactria, ungi, and plants. Viruses can also carry DNA between organisms, allowing transfer of genes even across biological domains. Large-scale gene transfer has also occurred between the ancestors of eukaryotic cells and prokaryotes, during the acquisition of chloroplasts and mitochondria.

Gene flow is the transfer of alleles from one population to another.

Migration into or out of a population may be responsible for a marked change in allele frequencies. Immigration may also result in the addition of new genetic variants to the established gene pool of a particular species or population. There are a number of factors that affect the rate of gene flow between different populations. One of the most significant factors is mobility, as greater mobility of an individual tends to give it greater migratory potential. Animals tend to be more mobile than plants, although pollen and seeds may be carried great distances by animals or wind.

Maintained gene flow between two populations can also lead to a combination of the two gene pools, reducing the genetic variation between the two groups. It is for this reason that gene flow strongly acts against speciation, by recombining the gene pools of the groups, and thus, repairing the developing differences in genetic variation that would have led to full speciation and creation of daughter species.

For example, if a species of grass grows on both sides of a highway, pollen is likely to be transported from one side to the other and vice versa. If this pollen is able to fertilise the plant where it ends up and produce viable offspring, then the alleles in the pollen have effectively been able to move from the population on one side of the highway to the other.

Genetic Structure

Because of physical barriers to migration, along with limited vagility, ad natal philopatry, natural populations are rarely panmictic

(Buston *et al.*, 2007). There is usually a geographic range within which individuals are more closely related to one another than those randomly selected from the general population. This is described as the extent to which a population is genetically structured (Repaci *et al.*, 2007). Genetic structuring can be caused by migration due to historical climate change, species range expansion or current availability of habitat.

Microbial Population Genetics

Microbial population genetics is a rapidly advancing field of investigation with relevance to many other theoretical and applied areas of scientific investigations. The population genetics of microorganisms lays the foundations for tracking the origin and evolution of antibiotic resistance and deadly infectious pathogens. Population genetics of microorganisms is also an essential factor for devising strategies for the conservation and better utilization of beneficial microbes (Xu, 2010).

History

Population Genetics: *Population genetics* was developed as a reconciliation of the Mendelian and biometrician models. A key step was the work of the British biologist and statistician R.A. Fisher. In a series of papers starting in 1918 and culminating in his 1930 book *The Genetical Theory of Natural Selection,* Fisher showed that the continuous variation measured by the biometricians could be produced by the combined action of many discrete genes, and that natural selection could change gene frequencies in a population, resulting in evolution (though lacking the knowledge of what an actual gene was at this time, it should be said in this sense he understood phenotypic trait frequency, rather than specifically identifiable gene frequency).

In a series of papers beginning in 1924, another British geneticist, J.B.S. Haldane, applied statistical analysis to real-world examples of natural selection, such as the evolution of industrial melanism in peppered moths, and showed that natural selection worked at an even faster rate than Fisher assumed.

The American biologist Sewall Wright, who had a background in animal breeding experiments, focused on combinations of interacting genes, and the effects of inbreeding on small, relatively isolated populations that exhibited genetic drift. In 1932, Wright introduced the concept of an adaptive landscape and argued that genetic drift and inbreeding could drive a small, isolated sub-population away from an adaptive peak, allowing natural selection to drive it towards different adaptive peaks. Fisher and Wright had some fundamental disagreements and a controversy about the relative roles of selection and drift continued for much of the century between the Americans and the British. The Frenchman Gustave Malécot was also important early in the development of the discipline.

The work of Fisher, Haldane and Wright founded the discipline of *population genetics*. This integrated natural selection with Mendelian genetics, which was the critical first step in developing a unified theory of how evolution worked.

John Maynard Smith was Haldane's pupil, whilst W.D. Hamilton was heavily influenced by the writings of Fisher. The American George R. Price worked with both Hamilton and Maynard Smith. American Richard Lewontin and Japanese Motoo Kimura were heavily influenced by Wright.

Modern Evolutionary Synthesis

In the first few decades of the 20th century, most field naturalists continued to believe that Lamarckian and orthogenic mechanisms of evolution provided the best explanation for the complexity they observed in the living world. However, as the field of genetics continued to develop, those views became less tenable. Theodosius Dobzhansky, a postdoctoral worker in T. H. Morgan's lab, had been influenced by the work on genetic diversity by Russian geneticists such as Sergei Chetverikov.

He helped to bridge the divide between the foundations of microevolution developed by the population geneticists and the patterns of macroevolution observed by field biologits, with his 1937 book *Genetics and the Origin of Species*. Dobzhansky examined the genetic diversity of wild populations and showed that, contrary to the assumptions of the population geneticists, these populations had large amounts of genetic diversity, with marked differences between sub-populations.

The book also took the highly mathematical work of the population geneticists and put it into a more accessible form. In Great Britain E.B. Ford, the pioneer of ecological genetics, continued throughout the 1930s and 1940s to demonstrate the power of selection due to ecologcal factors including the ability to maintain genetic diversity through genetic polymorphisms such as human blood types. Ford's work would contribute to a shift in emphasis during the course of the modern synthesis towards natural selection over genetic drift.

Genetic Analysis of Adenohypophysis Formation in Zebrafish

The Pituitary Gland (hypophysis) and the hypothalamus constitute an integrative center of endocrine control, linking the central nervous system and the endocrine system in regulating basic body functions and homeostasis. The pituitary itself comprises two anatomically and functionally distinct systems, the anterior lobe (adenohypophysis) and the posterior lobe (neurohypophysis).

The adenohypophysis is further subdivided into three regions: the pars anterior (or pars distalis), the pars tuberalis, and the pars intermedia. It contains at least six different cell types that are characterized by the different hormones they produce and secrete: lactotropes generating prolactin (PRL), somatotropes generating GH, thyrotropes generating TSH, gonatotropes generating FSH and LH, corticotropes generating ACTH, and melanotropes generating MSH. ACTH and MSH are formed via proteolytic cleavage from a common proprotein, proopiomelanocortin (POMC), which also gives rise to endorphins secreted by specific cells of the brain. MSH is generated in the intermediate lobe (pars intermedia) of the pituitary, which is rather rudimentary in human.

In contrast to the adenohypophysis, the neurohypophysis itself does not contain endocrine cells. Rather, it secretes hormones from axonal termini of hypothalamatic neuroendocrine cells that have innervated the neurohypophysis. These hormones are oxytocin (isotocin in fish) and vasopressin (vasotocin in fish), generated in the supraoptic and paraventricular nuclei of the hypothalamus. In addition, neuroendocrine cells from these and other hypothalamic nuclei elaborate peptide hormones (so called releasing or release inhibiting hormones), which control the activity of endocrine cells of the adenohypophysis. In contrast to the neurohypophyseal hormones, these hormones do not reach the pituitary via nerve fibers. Rather, they travel in the blood via the hypothalamic-hypophyseal portal system.

The development of the adenohypophysis and the formation of its different cell types from a common primordium is controlled both by intrinsic and extrinsic factors. The adenohypophysis is derived from placodal ectoderm at the anterior neural ridge, that becomes committed toward an adenohypophyseal fate via inductive signals from the ventral diencephalon, a subdivision of the forebrain that later gives rise to hypothalamus, infundibulum and neurohypophysis. During further development, the adenohypophyseal anlage remains under the control of signals from the hypothalamus. These signals, together with intrinsic signals of the adenohypophysis itself, regulate the maintenance, proliferation and differential specification of adenohypophyseal cells by sequential activation of different transcription factors, such as the Lim class homeodomain proteins Lhx3 or Lhx4, the paired-like homeodomain protein Prophet-of-Pit1, the Pou domain protein Pit1, the T-box factor Tpit/Tbx19, the zinc finger protein Gata2, and several others.

Final evidence for the requirement of such factors for pituitary development requires genetic analyses via loss-of-function mutants. In mouse, such mutants have been generated using gene knockout technology, as in the case of Lhx3, that thereby was shown to be required for all adenohypophyseal cell types except the corticotropes. More recently, conditional knockout techniques allow one to investigate the pituitary-specific role of genes also involved in other processes, as in the case of the steroidogenic factor Sf1, that is required downstream of Gata for the formation of gonadotropes.

However, such gene targeting approaches are generally biased because they require a priori knowledge of the molecular nature of the genes to be analyzed. As an alternative to this reverse genetics approach, genes essential for mouse pituitary development were isolated via positional cloning of spontaneous, viable mutations, such as Prop1, that is mutated in Ames dwarf mutants and required for somatotropes, lactotropes, thyrotropes, and gonadotropes, and Pit1, which is mutated in Snell and Jackson dwarf mutants and required for somatotropes, lactotropes and thyrotropes, defining the so-called Pit1 lineage of the adenohypophysis. Tpit/Tbx19 requirement for the Pomc lineage (corticotropes and melanotropes) was revealed via the analysis of human patients with isolated Acth deficiencies caused by recessive Tbx19 point mutations. However, as in other known human pituitary disorders, there are too few families and patients to allow a direct positional cloning of the affected human gene. Furthermore, the spontaneous mutation rates in humans and mice are too low to allow a saturating identification of all genes required for adenohypophysis development via forward genetics.

In light of these limitations of reverse and forward genetics in mammalian systems, we carried out a forward genetic analysis of pituitary mutants in a nonmammalian vertebrate, the zebrafish. Due to its extracorporal and rapid development, the transparency of its embryos and larvae, its high fecundity, its relatively small size and the ease of high-density maintenance, the zebrafish is highly suitable for mutant screens. In the past, three independent large-scale screens have been carried out, two of which used the chemical N-ethyl-N-nitrosourea (ENU) to introduce random point mutations over the entire genome.

Whereas in these screens mutant analyses were largely restricted to examining embryonic and larval morphology at different developmental stages, the more recent Tuebingen 2000 large-scale ENU screen was set up to allow mutant screening with molecular tools. As part of the Tuebingen 2000 screen, we searched for genes required for zebrafish adenohypophysis formation and patterning, carrying out large-scale whole mount *in situ* hybridizations with a probe detecting Gh encoding transcripts, a marker for somatotropes. The staining was necessary because the adenohypophysis of zebrafish larvae is morphologically too indistinct to be analyzed in large numbers without molecular tools.

Recent work has revealed both crucial similarities and differences in the development of the adenohypophysis between fish and mammals. The zebrafish pituitary contains the same cell types as the mammalian pituitary, however, there are crucial differences in the morphogenesis and the architecture of the glands in the different vertebrate species. Thus, the zebrafish adenohypophysis maintains its placodal organization and remains in a subepithelial position after oral cavity formation, whereas no invagination of the oral ectoderm equivalent to Rathke's pouch formation in mammals takes place.

Furthermore, there are differences in the patterning of the adenohypophyseal anlage: in zebrafish, the different cell types are distributed in three distinct domains along the antero-posterior axis of the pituitary, rather than along the dorsoventral axis as in mammals, with the Msh-generating pars intermedia most posteriorly, and the neurohypophysis located dorsal of adenohypophysis. In addition, fish and mammals display significant differences in the onset of the specification of the lactotrope lineage, that is the first lineage to specify in fish, but the last in mouse. Despite such differences, crucial mechanisms of adenohypophyseal induction and patterning appear to be highly conserved among fish and mammals, as for instance indicated by the conserved role of Sonic Hedgehog signalling from the ventral diencephalon. In light of these data, the large-scale ENU mutagenesis screen was expected to uncover genes regulating fish-specific features of the adenohypophysis, as well as genes with shared functions in mouse and fish. This would further illuminate the degree of conservation between the two vertebrate species, and possibly identify novel regulators also involved in mammalian pituitary development.

Results

Identification of Mutations Affecting Zebrafish Adenohypophysis Development: As part of a large-scale diploid F3 ENU mutagenesis screen to uncover recessive zygotic effect mutations, we screened for mutations affecting zebrafish adenohypophysis development. In total, F3 clutches of 4584 F2 families, representing 4253 mutagenized haploid genomes, were screened at 120 h post fertilization via whole mount *in situ* hybridization with a probe detecting transcripts of *gh*, a marker of the somatotropes.

The late developmental stage for mutant screening was chosen for two reasons: 1) to reveal also defects in later steps of pituitary development or patterning that would be missed when mutants were investigated much earlier to avoid pituitary phenotypes that are secondary consequences of early and general defects, such as early brain necrosis, that occurs very frequently, but usually leads to embryonic death between 24 and 72 hpf.

Secondary consequences on pituitary formation have also been revealed for midline mutants defective in signalling by Nodal and Hedgehog family members. In such mutants, eye field separation and hypothalamus formation are compromised, leading to pituitary defects caused by the lack of inductive signalling from the hypothalamus, consistent with the loss of the pituitary in hypothalamus-deficient mouse embryos. However, in contrast to pituitary-specific mutants, such zebrafish midline null mutants usually die before 120 hpf, although secondary pituitary defects caused by more subtle, viable midline mutations cannot be ruled out.

In total, we identified 13 mutants that were viable at 120 hpf but displayed a severe reduction or complete absence of *gh* staining in whole mount *in situ* hybridization. Other gross abnormalities, such as brain degeneration, which could have secondarily affected the pituitary, were not detected in these mutants. Only two of the 13 mutants (allele numbers t24594 and t20626) showed reduced eye distance and a curled-down tail, indicative for midline defects. Although t20626 was not further investigated, genomic mapping placed t24594 31–32 centimorgans from the top of linkage group 6, just below the marker Z6626. This is the same position as reported for the midline mutation *iguana*, and strongly suggests that t24594 is a weak *iguana* allele that we did not analyze further.

Of the remaining 11 adenohypophysis-specific mutations, eight were recovered. A combination of complementation crossing and genomic mapping revealed that they fall into four complementation groups, defining four genes essential for adenohypophysis development in the zebrafish. The genomic positions of the four genes. All genes appear to be dispensable for hypothalamus development, as revealed by the normal expression of *nkx2.1* in mutant embryos at 32 hpf, and the normal number of *isotocin*-positive cells at 32 hpf. *isotocin* cDNA was isolated in our laboratory via degenerate RT-PCR, and is identical with the cDNA recently reported by Unger and Glasgow. It is expressed in two nuclei of magnocellular neurons in the anterior hypothalamus, possibly the presumptive supraoptic or paraventricular nuclei, from where the peptide hormone is supposed to undergo axonal transport into the more ventrally and posteriorly located region of the neurohypophysis adjacent to the adenohypophysis.

In contrast to the normally developing hypothalamus and neurohypophysis, mutants in all four genes show specific, but different phenotypes in the adenohypophysis. Two of the mutants, *lim absent* (*lia*), represented by four alleles, and *pituitary absent* (*pia*), represented by one allele, lack the entire adenohypophysis. Whereas in *pit1* mutants, represented by two alleles, and in *all-absent-except-lactotropes* (*aal*) mutants, represented by one allele, only some of the pituitary cell types are absent.

Mutations Affecting the Entire Adenohypophysis: As in mammals and birds, the adenohypophysis of the zebrafish embryo is supposed to be of placodal origin, derived from medial cells of the anterior neural ridge, and laterally flanked by cells of the olfactory placodes. Whereas a few markers like *anf* are expressed in the anterior neural ridge at late gastrula and early segmentation stages, none continues to be expressed until cells can be identified as adenohypophyseal. In addition, none of them is altered in our mutants. In this respect, the currently known earliest *bona fide* marker for the presumptive zebrafish adenohypophysis is *lim3*, the homolog of *Lhx3* required for Rathke's pouch formation in mouse. In zebrafish, *lim3* starts to be expressed at the 21-somite stage, presumably in all cells of the pituitary anlage, when it is still organized in a placodal fashion in front of the forming head.

Differentiation of the first two pituitary cell lineages, the *prl* expressing lactotropes and the *pomc* expressing corticotropes and/or melanotropes, occurs very soon after the onset of *lim3* expression in lateral regions of the *lim3* domain, whereas expression of *tsh* in thryotropes and *gh* in somatotropes only starts later, after the pituitary has moved inwards. At 72 hpf, *prl*, *pomc*, *tsh*, and *gh* are expressed in distinct, but partially overlapping domains along the antero-posterior axis of the adenohypophysis.

lia and *pia* mutants lack the expression of all four adenohypophysis hormones at 72 hpf and 120 hpf. The remaining expression of *pomc* in two longitudinal stripes is confined to ß-endorphin-synthesizing cells in the ventral base of the diencephalon, whereas melanotropes and corticotropes of the adenohypophysis, as well as lactotropes, thyrotropes, and somatotropes, are absent.

In contrast to their indistinguishable phenotypes at d 3 and 5 after fertilization, *lia* and *pia* mutants differ in *lim3* expression during earlier stages of development. Although in *lia* mutants, *lim3* expression in the region of the adenohypophysis anlage is completely absent at 32 hpf, *lim3* expression in *pia* mutants is reduced, but clearly present. However, *prl* expression at 32 hpf is absent in both mutants. In sum, the expression pattern analyses suggest that both *lia* and *pia* are required for the specification of all adenohypophyseal cell types, with *lia* acting earlier than *pia*. *lia* appears to act upstream of *lim3*, whereas *pia* is more likely to act at the level, in parallel or downstream of *lim3*. The phenotype of *pia* zebrafish mutants is quite similar to that of *Lhx3/Lhx4* double mutant mice, suggesting that it might be caused by a mutation in a zebrafish *lim* gene. However, *lim3* itself can be ruled out, as it maps to a different linkage group.

Mutations Affecting Adenohypophyseal Patterning and Lineage Specification: The different cell types of the adenohypophysis derive from a common primordium, the placode, that initially consists of identical precursor cells. Subsequently, the placode becomes patterned, and cells in different regions of the primordium undergo differential specifications to form the different cell types characterized by the hormones they produce and secrete. Data from mammalian pituitary development suggest that the patterning process is characterized by subdivision of the common pool of precursor cells into certain cell lineages that then continue to branch until single cell types are generated. It is assumed that cells of the Pomc lineage, giving rise to corticotropes and melanotropes, split off first. In contrast to all other pituitary cell types, they express and require the T-box transcription factor Tpit/Tbx19, but are independent of the Lim class homeodomain transcription factor Lhx3. The complementary lineage, consisting of thyrotropes, somatotropes, lactotropes, and gonadotropes, is characterized by its Lhx3 dependence. Another lineage, comprising a subset of the Lhx3-dependent cell types, is defined via its dependence on the Pou domain transcription factor Pit1, and therefore called the Pit1 lineage. It consists of thyrotropes, lactotropes, and somatotropes, all of which are absent in *Pit1* mutant mice.

In our zebrafish screen, we could identify two genes required for the specification of two different adenohypophyseal cell lineages. Mutants in both genes show normal *lim3* expression at 32 hpf, indicating that the pituitary anlage is induced normally and of normal size. However, already at this early stage (32 hpf), the subdivision within the anlage is altered, indicated by the loss of *prl* expression in *pit1* mutants, whereas *prl* expression in *aal* mutants appears normal.

At 120 hpf, *pit1* mutant embryos are characterized by loss of lactotropes, somatotropes and thyrotropes, whereas *pomc*-positive adenohypophyseal cells, most likely anterior corticotropes and posterior melanotropes, are present at normal or even slightly elevated numbers. The zebrafish *pit1* mutants resemble Pit1-deficient mice, and indeed could be shown to carry mutations in the zebrafish *pit1* gene. *aal* mutant zebrafish show a very different phenotype; they fail to generate *pomc*-, *tsh*-, and *gh*-expressing pituitary cells, whereas the *prl*-expressing lactotropes are still present both at 32 hpf and 120 hpf. However, at the later developmental stages, the lactotropes are dispersed along the antero-posterior axis, rather than being organized in the sharp anterior domain found in wild-type siblings. This phenotype seems to define a new lineage of adenohypophyseal cells, consisting of corticotropes, melanotropes, thyrotropes, and somatotropes, all of which depend on the *aal* gene, whereas the lactotropes appear to develop independently of *aal* function. In contrast to *pit1*, no phenotype similar to that of *aal* mutants has been described in mouse thus far. Thus, the *aal* mutation seems to have revealed a thus far unknown lineage of adenohypophyseal cells that partially overlaps with the Pit1 lineage.

By screening approximately 1.5 times more mutagenized haploid genomes than in the first Tuebingen large-scale screen, we isolated eleven mutants with specific defects in the formation and patterning of the zebrafish adenohypophysis. Of those 11, three could not be recovered as yet due to logistic problems. The remaining eight mutations fall into four complementation groups, defining four genes named *lia*, *pia*, *pit1*, and *aal*, all of which appear to be required for different steps of adenohypophysis development.

Genes Required Upstream and Downstream of Lim3 to Drive Development of All Adenohypophyseal Cell Types

Two genes, *lia* and *pia*, are required for all adenohypophyseal cell types. In contrast, other placodal derivatives appear to develop normally in mutant embryos, as for instance indicated by the normal expression of marker genes of the olfactory epithelium (Herzog, W., and M. Hammerschmidt, unpublished data), which derives from placodal positions located left and right of the adenohypophyseal placode at the anterior neural ridge.

This indicates that *lia* and *pia* specifically act on early steps of adenohypophyseal placode formation or maintenance, whereas other placodes are regulated differently. During adenohypophyseal development, *lia* and *pia* appear to be required for different, maybe subsequent steps, as indicated by the differences in the *lim3* expression in the two mutants. Thus, *lia* appears to act at an earlier step, before *lim3* expression in the adenohypophyseal placode is initiated, whereas *pia* acts after the initiation of *lim3* expression. However, this does not necessarily mean that *lia* fulfills its indispensable function via the activation of *lim3*, nor does it necessarily mean that *lim3* is required for *pia* activation. *lia* could as well activate other essential genes that act in parallel or in addition to *lim3*, and *pia* could as well act in parallel rather than downstream of *lim3*. However, it is interesting to note that *pia* mutants—despite the early presence of *lim3* transcripts—fail to express even early adenohypophyseal hormone genes like *prl*. This indicates that in the absence of *pia*, *lim3* is not sufficient for early lactotrope specification—although it appears to be necessary. The latter is suggested by our preliminary results obtained with antisense morpholino oligonucleotides, according to which inactivation of *lim3* leads to the loss of all adenohypophyseal cell lineages except some *pomc* cells. Final analyses of the epistatic relationships between *lia*, *lim3*, and *pia* will only be possible after the *lia* and *pia* genes have been cloned.

To gain further insight into the biological roles of *lia* and *pia* during the development of the zebrafish adenohypophysis, future experiments must reveal the fate of the pituitary precursor cells in the mutants. Thus, cell-tracing experiments, bromodeoxyuridine incorporation studies, and transferase deoxyuridine triphosphate nick-end labelling or acridine orange stainings will be carried out to investigate whether adenohypophyseal cells are lost because of transfating, failed proliferation, or cell death. In addition, analyses of chimeric embryos generated via cell transplantation will help to specify in which cell types the two genes are required.

Genes Required for Lineage Specification: Evidence for a Conserved and a Novel Lineage of Adenohypophyseal Cell Types

In contrast to *lia* and *pia*, *pit1* and *aal* mutants lose only some of the adenohypophyseal cell types, indicating that they are required for adenohypophyseal patterning and lineage specification processes during later stages of pituitary development. *pit1* mutants lack *gh*, *prl*, and *tsh* staining, whereas *pomc* staining is normal or even enlarged, particularly in its posterior domain that accommodates both corticotropes and melanotropes.

The phenotype looks very similar to that of mouse *Pit1* mutants, characterized by the loss of thyrotropes, somatotropes, and lactotropes. Due to the lack of suitable markers, we could not study the effect of zebrafish *pit1* on the gonadotropes, that according to the mouse mutant should be present in *pit1* mutant fish. One subtle difference in the phenotypes of mouse and zebrafish mutants is that zebrafish mutants lack all *tsh* expression, whereas in mouse mutants, only the caudo-medial thyrotropes are lost, although the earlier specifying rostral-tip thyrotropes are present. *Tsh* expression in this special thyrotrope lineage is independent of Pit1, and most likely regulated by the leucine zipper transcription factor TEF. The functional importance of the rostral-tip thyrotrope cells is not clear. However, our data obtained for the zebrafish *pit1* mutant suggest that they might be a specialty of mammals.

In contrast to *pit1*, the zebrafish *aal* mutants appear to define a thus far unknown lineage of adenohypophyseal cell types, lacking somatotropes, thyrotropes, corticotropes, and melanotropes, but not the lactotropes. Such a combination of lost cell types had not been previously described. In addition, the parallel existence of the Pit1 and Aal lineages indicates mechanisms of lineage specification beyond the thus far believed subsequent branching off of lineages from a common precursor pool.

Consistent with such a linear mechanism, only two relative patterns of cell lineages had been observed thus far. Lineages were either complementary (such as the Pomc lineage and the Prop1 lineage), or part of each other (such as Pit1, consisting of somatotropes, lactotropes, and thyrotropes, and Prop1, consisting of the Pit lineage plus the gonadotropes).

In contrast, the zebrafish Pit1 and Aal lineages show a novel relative pattern, with shared cell types, as well as cell types specific for one or the other lineage. Thus, somatotropes and thyrotropes appear to belong both to the Pit1 and the Aal lineage. However, corticotropes and melanotropes only belong to the Aal lineage and are independent of Pit1, whereas the lactotropes belong to the Pit1 lineage only and are independent of Aal. This indicates that factors driving cell lineage specification are not only used in a mutually exclusive or consecutive fashion, but can also be recruited in different combinations to allow differential parallel cell specifications.

In mammals, no Aal lineage has been identified as yet. However, this does not necessarily mean that the mammalian Aal homolog is not required or involved in adenohypophysis development. Rather, the exclusive regulation of *prl* expression independently of Aal function might be a specialty of fish, consistent with the much earlier onset of *prl* expression in fish compared with mouse, and with the earlier and additional function of Prl during osmoregulation in water-living larvae. Along these lines, it is tempting to speculate that during the evolution of water-living vertebrates, lactotropes became independent of Aal, or that during evolution of land-based vertebrates, lactotropes got under the control of the Aal homolog, similar to the other pituitary cell types. In this case, mutations in the Aal homolog in mouse would affect all adenohypophyseal cell lineages. Final answers have to await the cloning of the zebrafish *aal* gene, and the identification and analysis of possible mammalian homologues.

The Cloning of the Mutated Pituitary Genes

Because the introduction of point mutations during the ENU mutagenesis is random, the molecular nature of the genes affected in the isolated mutants is not known a priori, but has to be determined in subsequent steps. However, with the ease of mutant embryo collection, together with recent progress in the zebrafish genome projects, positional cloning of such ENU-mutated genes has become relatively easy, and has been successfully applied in multiple cases. Using a combination of positional cloning and candidate testing, we have for instance been able to identify the molecular nature of our two zebrafish *pit1* alleles. Also, future cloning of *lia*, *pia*, and *aal* will hopefully identify novel genes in pituitary development. In addition, we will be able to investigate whether such genes are specifically required for pituitary development in fish, or whether their roles are conserved between fish and mammals.

Possible Reasons for the Low Number of Identified Pituitary Genes and Perspectives

Based on gene targeting and cloning of spontaneous mutations, a total of approximately 20 genes indispensable for pituitary development in the mouse were identified. These include genes required for the combined formation of hypothalamus, neurohypophysis, and adenohypophysis, genes required for hypothalamus and neurohypophysis only, and genes required for the adenohypophysis only. Our screen was designed to find adenohypophysis-specific defects only. Pituitary phenotypes caused secondarily due to loss of the hypothalamus would have been missed because screening was performed at a late developmental stage, after mutations in such genes should have been lethal.

In addition, the zebrafish mutants were screened for altered *gh* expression only—the only adenohypophyseal hormone gene that had been cloned at the time the screen was carried out. The *gh* probe should have allowed us to identify genes required for adenohypophysis induction and development in general, as well as genes specifically required for somatotropes and somatotropes-including lineages, such as *Lhx3*, *Prop1*, and *Pit1*. On one hand, *gh* was a good choice, because it allowed us to identify genes that would most likely have been missed with other hormone probes (*pit1* mutants still express *pomc*; *aal* mutants still express *prl*). On the other hand, we might have missed genes that would have been revealed with other probes, such as *Tpit/*

Tbx19 (affecting the Pomc lineage only), *Gata2* (affecting gonatotropes and thyrotropes only), or *Sf1* (affecting gonadotropes only). In light of these possible limitations of the *gh* probe, we are currently preparing another large-scale ENU screen, looking for the expression of *pit1*, *lim3*, *isotocin*, *pomc*, and *prl* at early and late stages of zebrafish development.

Another reason for the low number of identified essential zebrafish genes could be the larger size and complexity of the zebrafish genome. Due to the additional genome duplication that has occurred during teleost evolution, genes controlling pituitary development in zebrafish might display a higher degree of functional redundancy than in mammals. However, genome analyses suggest that only approximately 25% of the genes gained in the duplication have remained active, with most of them having evolved a different expression pattern than their paralogs. Also, parallel searches for zebrafish genes required for other developmental processes, such as angiogenesis and thymus development (Thomas Boehm, Max-Planck Institute for Immunobiology, Freiburg, Germany; personal communication) have yielded many more mutants, suggesting that high functional redundancy with a low number of indispensable genes might be a feature of some, but not all processes of zebrafish development. On the other hand, we cannot rule out general technical problems during our mutagenesis. Comparing our mutagenesis conditions with those of previous screens, the screening of over 4200 mutagenized genomes should have yielded 50–80% saturation, with an average allele frequency per gene between three and four. However, in case of our identified pituitary genes, the average allele rate was two, with single alleles for two of the four genes. This suggests that the ENU mutagenesis might have been less efficient than in previous screens. Further complementation testing of other mutant classes has to be carried out for final conclusion about the mutation rates.

Mapping

Genomic localization of zebrafish mutations was performed as described, using the Tübingen marker set for genome scans (version 4) on F2 Tuebingen x Wik crosses of the mutant carriers. Primer sequences are available from the Massachusetts General Hospital web site.

Cloning of the Zebrafish Isotocin cDNA

Isotocin cDNA was cloned via degenerate RT-PCR, using the following primers: sense 5'TGY TAY ATH CAR AAY TGY CC, antisense

5'CCR CAR CAD ATN BWN GGN CC, with 35 cycles and an annealing temperature of 53 C. To obtain the full-length cDNA sequence, this was followed by a 3'-rapid amplification of cDNA ends (RACE), using the sense primer and the Smart Race Kit (Clontech, Palo Alto, CA) according to the manufacturer's instructions. PCR fragments were cloned into pCRII (Invitrogen, Carlsbad, CA). For *isotocin in situ* antisense probe synthesis, the plasmid was digested with *Kpn*I and transcribed using T7 RNA polymerase.

In Situ Hybridizations

For initial screening, F3 clutches were incubated in the presence of 0.25 mM 1-phenyl-2-thiourea (Sigma, St. Louis, MO) to avoid melanin synthesis, and fixed in 4% paraformaldehyde/PBS at 120 hpf. Whole mount *in situ* hybridization was carried out in specially designed 48-well plates (Aldinger, Nagold, Germany) with digoxygenin-labeled probes for *gh*, and *rag1*, a marker for thymic T cells, after standard protocols. Red/blue double *in situ* hybridizations were carried out as described. *prl*, *pomc*, *tsh*, *gh*, *lim3*, and *nkx2.1* probes were synthesized as reported.

Dual Inheritance Theory

Dual inheritance theory (DIT), also known as gene-culture coevolution, was developed in the late 1970s and early 1980s to explain how human behaviour is a product of two different and interacting evolutionary processes: genetic evolution and cutural evolution. DIT is a "middle-ground" between much of social science, which views culture as the primary cause of human behavioural variation, and human sociobiology and evolutionary psychology which view culture as an insignificant by-product of genetic selection. In DIT, culture is defined as information in human brains that got there by social learning. Cultural evolution is considered a Darwinian selection process that acts on cultural information. Dual Inheritance Theorists often describe this by analogy to genetic evolution, which is a Darwinian selection process acting on genetic information.

Because genetic evolution is relatively well understood, most of DIT examines cultural evolution and the interactions between cultural evolution and genetic evolution.

Theoretical Basis

DIT holds that genetic and cultural evolution interact in the evolution of *Homo sapiens*. DIT recognizes that the natural selection of genotypes is an important component of the evolution of human behaviour and that cultural traits can be constrained by genetic imperatives. However, DIT also recognizes that genetic evolution has endowed the human species with a parallel evolutionary process of cultural evolution. DIT makes three main claims:

Culture Capacities are Adaptations: The human capacity to store and transmit culture arose from genetically evolved psychological mechanisms. This implies that at some point during the evolution of the human species a type of social learning leading to cumulative cultural evolution was evolutionarily advantageous.

Culture Evolves: Social learning processes give rise to cultural evolution. Cultural traits are transmitted differently than genetic traits and, therefore, result in different population-level effects. These effects can help explain human behavioural variation.

Genes and Culture Coevolve: Cultural traits alter the social and physical environments under which genetic selection operates. For example, the cultural adoptions of agriculture and dairying have, in humans, caused genetic selection for the traits to dgest starch and lactose, respectively. As anothr example, it is likely that once culture became adaptive, genetic selection caused a refinement of the cognitive architecture that stores and transmits cultural information. This refinement may have further influenced the way culture is stored and the biases that govern its transmission.

DIT also predicts that, under certain situations, cultural evolution may select for traits that are genetically maladaptive. An example of this is the demographic transition, which describes the fall of birth rates within industrialized societies. Dual Inheritance Theorists hypothesize that the demographic transition may be a result of a prestige bias, where individuals that forgo reproduction to gain more influence in industrial societies are more likely to be chosen as cultural models.

View of Culture

People have defined the word "culture" to describe a large set of different phenomena. A definition that sums up what is meant by "culture" in DIT is:

Culture is information stored in individuals' brains that is capable of affecting behaviour and that got there through social learning.

This view of culture emphasizes population thinking by focusing on the process by which culture is generated and maintained. It also views culture as a dynamic property of individuals, as opposed to the more standard social science view of culture as a superorganic entity to which individuals must conform. This view's main advantage is that it connects individual-level processes to population-level outcomes.

Genetic Influence on Cultural Evolution

Genes have an impact on cultural evolution via psychological predispositions on cultural learning. Genes encode much of the information needed to form the human brain. Genes constrain the brain's structure and, hence, the bility of the brain to acquire and store culture.Genes may also endow individuals with certain types of transmission bias (described below).

Cultural Influences on Genetic Evolution

Culture can profoundly influence gene frequencies in a population. One of the best known examples is the prevalence of the genotype for adult lactose absorption in human populations, such as Northern Europeans and some African societies, with a long history of raising cattle for milk. Other societies such as East Asians and Amerindians, retain the typical mammalian genotype in which the body shuts down lactase production shortly after the normal age of weaning. This implies that the cultural practice of raising cattle for milk led to a selection for genetic traits for lactose digestion. Recently, analysis of natural selection on the human genome suggests that civilization has accelerated genetic change in humans over the past 10,000 years.

Mechanisms of Cultural Evolution

In DIT, the evolution and maintenance of cultures is described by five major mechanisms, natural selection of cultural variants, random variation, cultural drift, guided variation and transmission bias.

Natural Selection: Cultural differences among individuals can lead to differential survival of individuals, with those that survive more likely to have their cultural variants adopted by others. The patterns of this selective process depend on transmission biases and can result in behaviour that is more adaptive to a given environment.

Random Variation: Random variation arises from errors in the learning, display or recall of cultural information.

Cultural Drift: Cultural drift is a process roughly analogous to genetic drift in evolutionary biology. In cultural drift, the frequency of cultural traits in a population may be subject to random fluctuations due to chance variations in which traits are observed and transmitted (sometimes called "sampling error"). These fluctuations might cause cultural variants to disappear from a population. This effect should be especially strong in small populations.

An example of cultural drift would be dialects of song birds. If a group of songbirds are culturally split (viz, no interactions with their songs), then distinct dialects will form from the original mating songs of the male birds. This is thought to happen due to errors in songbird singing and acquisition by successive generations, which, while still functional, do not maintain fidelity with the original song of the first generation.

Guided Variation: Cultural traits may be gained in a population through the process of individual learning. Once an individual learns a novel trait, it can be transmitted to other members of the population. The process of guided variation depends on an adaptive standard that determines what cultural variants are learned.

Biased Transmission: Understanding the different ways that culture traits can be transmitted between individuals has been an important part of DIT research since the 1970s. Transmission biases occur when some cultural variants are favoured over others during the process of cultural transmission. Boyd and Richerson (1985) defined and analytically modeled a number of possible transmission biases. The list of biases has been refined over the years, especially by Henrich and McElreath.

Content Bias: Content biases result from situations where some aspect of a cultural variant's content makes them more likely to be adopted. Content biases can result from genetic preferences, preferences determined by existing cultural traits, or a combination of the two. For example, food preferences can result from genetic preferences for sugary or fatty foods and socially-learned eating practices and taboos. Content biases are sometimes called "direct biases."

Context Bias: Context biases result from individuals using clues about the social structure of heir population to determine what cultural variants to adopt. This determination is made without reference to the content of the variant. There are two major categories of context biases: (1) model-based biases, and (2) frequency-dependent biases.

Model-Based Biases: Model-based biases result when an individual is biased to choose a particular "cultural model" to imitate. There are four major categories of model-based biases: (1) prestige bias, (2) skill bias, (3) success bias, (4) similarity bias. A "prestige bias" results when individuals are more likely to imitate cultural models that are seen as having more prestige. A measure of prestige could be the amount of deference shown to a potential cultural model by other individuals. A "skill bias" results when individuals can directly observe different cultural models performing a learned skill and are more likely to imitate cultural models that perform better at the specific skill.

A "success bias" results from individuals preferentially imitating cultural models that they determine are most generally successful (as opposed to successful at a specific skill as in the skill bias.) A "similarity bias" results when individuals are more likely to imitate cultural models that are perceived as being similar to the individual based on specific traits.

Frequency-Dependent Biases: Frequency-dependent biases result when an individual is biased to choose particular cultural variants based on their perceived frequency in the population. The most explored frequency-dependent bias is the "conformity bias." Conformity biases result when individuals attempt to copy the mean or the mode cultural variant in the population. Another possible frequency dependent bias is the "rarity bias." The rarity bias resuls when individuals preferentially choose cultural variants that are less common in the population. The rarity bias is also sometimes called a "nonconformist bias".

Social Learning and Cumulative Cultural Evolution

In DIT, the evolution of culture is dependent on the evolution of social learning. Analytic models show that social learning becomes evolutionarily beneficial when the environment changes with enough frequency that genetic inheritance can not track the changes, but not fast enough that individual learning is more efficient. While other species have social learning, and thus some level of culture, only humans, some birds and chimpanzees are known to have cumulative culture. Boyd and Richerson argue that the evolution of cumulative culture depends on observational learning and is uncommon in other species because it is ineffective when it is rare in a population. They propose that the environmental changes occurring in the Pleistocene may have provided the right environmental conditions. Michael Tomasello argues that cumulative cultural evolution results from a "ratchet effect" that began when humans developed the cognitive architecture to understand others as mental agents. Furthermore Tomasello proposed in the 80s that there are some disparities between the observational learning mechanisms found in humans and great apes - which go some way to explain the observable difference between great ape traditions and human types of culture.

Cultural Group Selection

Although group selection is commonly thought to be nonexistent or unimportant in genetic evolution, DIT predicts that, due to the nature of cultural inheritance, it may be an important force in cultural evolution. The reason group selection is thought to operate in cultural evolution is because of conformist biases. Conformist biases make it difficult for novel cultural traits to spread through a population. Conformist bias also helps maintain variation between groups. These two properties, rare in genetic transmission, are necessary for group selection to operate. Based on an earlier model by Cavalli-Sforza and Feldman, Boyd and Richerson show that conformist biases are almost inevitable when traits spread through social learning, implying that group selection is common in cultural evolution. Analysis of small groups in New Guinea imply that cultural group selection might be a good explanation for slowly changing aspects of social structure, but not for rapidly changing fads. The ability of cultural evolution to maintain intergroup diversity is what allows for the study of cultural phylogenetics.

Historical Development

The idea that human cultures undergo a similar evolutionary process as genetic evolution goes back at least to Darwin In the 1960s, Donald T. Campbell published some of the first theoretical work that adapted principles of evolutionary theory to the evolution of cultures. In 1976, two developments in cultural evolutionary theory set the stage for DIT. In that year Richard Dawkins's *The Selfish Gene* introduced ideas of cultural evolution to a popular audience. Although one of the best-selling science books of all time, because of its lack of mathematical rigor, it had little impact on the development of DIT. Also in 1976, geneticists Marcus Feldman and Luigi Luca Cavalli-Sforza published the first dynamic models of gene-culture coevolution. These models were to form the basis for subsequent work on DIT, heralded by the publication of three seminal books in 1980 and 1981.

The first was Charles Lumsden and E.O. Wilson's *Genes, Mind and Culture*. This book outlined a series of mathematical models of how genetic evolution might favour the selection of cultural traits and how cultural traits might, in turn, affect the speed of genetic evolution. While it was the first book published describing how genes and culture might coevolve, it had relatively little impact on the further development of DIT. Some critics felt that their models depended too heavily on genetic mechanisms at the expense of cultural mechanisms. Controversy surrounding Wilson's sociobiological theories may also have decreased the lasting impact of this book.

The second 1981 book was Cavalli-Sforza and Feldman's *Cultural Transmission and Evolution: A Quantitative Approach.* Borrowing heavily from population genetics and epidemiology, this book built a mathematical theory concerning the spread of cultural traits. It describes the evolutionary implications of vertical transmission, passing cultural traits from parents to offspring; oblique transmission, passing cultural traits from any member of an older generation to a younger generation; and horizontal transmission, passing traits between members of the same population.

The next significant DIT publication was Robert Boyd and Peter Richerson's 1985 *Culture and the Evolutionary Process.* This book presents the now-standard mathematical models the evolution of social learning under different environmental conditions, the population effects of social learning, various forces of selection on cultural learning rules, different forms of biased transmission and their population-level effects, and conflicts between cultural and genetic evolution. The book's conclusion also outlined areas for future research that are still relevant today.

Current and Future Research

In their 1985 book, Boyd and Richerson outlined an agenda for future DIT research. This agenda, outlined below, called for the development of both theoretical models and empirical research. DIT has since built a rich tradition of theoretical models over the past two decades. However, there has not been a comparable level of empirical work.

In a 2006 interview Harvard biologist E.O. Wilson expressed disappointment at the little attention afforded to DIT:

> *"...for some reason I haven't fully fathomed, this most promising frontier of scientific research has attracted very few people and very little effort."*

Kevin Laland and Gillian Brown attribute this lack of attention to DIT's heavy reliance on formal modeling, which doesn't attract the media attention of less rigorous approaches to human behavioural evolution, such as evolutionary psychology:

> "In many ways the most complex and potentially rewarding of all approaches, [DIT], with its multiple processes and cerebral onslaught of sigmas and deltas, may appear too abstract to all but the most enthusiastic reader. Until such a time as the theoretical hieroglyphics can be translated into a respectable empirical science most observers will remain immune to its message."

Economist Herbert Gintis disagrees with this critique, citing existing empirical work as well as more recent work using techniques from behavioural economics. These behavioural economic techniques have been adapted to test predictions of cultural evolutionary models in laboratory settings as well as studying differences in cooperation in fifteen small-scale societies in the field.

Since one of the goals of DIT is to explain the distribution of human cultural traits, ethnographic and ethnologic techniques may also be useful for testing hypothesis stemming from DIT. Although findings from traditional ethnologic studies have been used to buttress DIT arguments, thus far there have been little ethnographic fieldwork designed to explicitly test these hypotheses. A major difficulty in using existing ethnographic data to test DIT is that, due to cultural anthropology's assumption of culture as a superorganic entity, ethnographic data tends to ignore individual and intragroup cultural variation and focus almost entirely on intergroup variation.

DIT has been viewed as having great potential for unifying diverse academic fields under one overarching theory. Mesoudi, et al. have identified DIT as the ideal way to build a comprehensive theory of cultural evolution to answer questions about human behaviour at different temporal and spacial scales. Along with game theory, Herb Gintis has named DIT one of the two major conceptual theories with potential for unifying the behavioural sciences, including economics, biology, anthropology, sociology, psychology and political science. Because it addresses both the genetic and cultural components of human inheritance, Gintis sees DIT models as providing the best explanations for the ultimate cause of human behaviour and the best paradigm for integrating those disciplines with evolutionary theory. In a review of competing evolutionary perspectives on human behaviour, Laland and Brown see DIT as the best candidate for uniting the other evolutionary perspectives under one theoretical umbrella.

Relation to other Fields

Sociology and Cultural Anthropology: Two major topics of study in both sociology and cultural anthropology are human cultures and cultural variation. However, Dual Inheritance theorists charge that both disciplines too often treat culture as a static superorganic entity that dictates human behaviour. Cultures are defined by a suite of common traits shared by a large group of people, without regard to variation in cultural traits at the individual level. This is in sharp contrast to DIT, which models human culture at the individual level and views culture as the result of a dynamic evolutionary process at the population level.

Human Sociobiology and Evolutionary Psychology: Human sociobiologists and evolutionary psychologists try to understand how maximizing genetic fitness, in either the modern era or past environments, can explain human behaviour. To the human sociobiologist and evolutionary psychologist, culture is either trivial or so bound by genetic fitness that it is unimportant. When faced with a common and seemingly maladaptive trait, practitioners from these disciplines try to determine how the trait actually increases genetic fitness (maybe through kin selection or by speculating about early evolutionary environments). Dual Inheritance theorists, in contrast, will consider a variety of genetic and cultural processes in addition to natural selection on genes.

Human Behavioural Ecology: Human behavioural ecology (HBE) and DIT have a similar relationship to what ecology and evolutionary biology have in the biological sciences. HBE is more concerned about ecological process and DIT more focused on historical process. One difference is that human behavioural ecologists often assume that culture is a system that produces the most adaptive outcome in a given environment. This implies that similar behavioural traditions should be found in similar environments. However, this is not always the case. A study of African cultures showed that cultural history was a better predictor of cultural traits than local ecological conditions.

Memetics: Memetcs, which comes from the meme idea described in Dawkins's *The Selfish Gene*, is similar to DIT in that it treats culture as an evolutionary process that is distinct from genetic transmission. However, there are some philosophical differences between memetics and DIT. One difference is that memetics' focus is on the selection potential of discrete replicators (memes), where DIT allows for transmission of both non-replicators and non-discrete cultural variants. DIT does not assume that replicators are necessary for cumulative adaptive evolution. DIT also more strongly emphasizes the role of genetic inheritance in shaping the capacity for cultural evolution. But perhaps the biggest difference is a difference in academic lineage. Memetics as a label is more influential in popular culture than in academia. Critics of memetics argue that it is lacking in empirical support or is conceptually ill-founded, and question whether there is hope for the memetic research programme succeeding. Proponents point out that many cultural traits are dscrete, and that many existing models of cultural inheritance assume discrete cultural units, and hence involve memes.

Bibliography

Aldridge, S.: *The Thread of Life: The Story of Genes and Genetic Engineering*, Cambridge University Press, Cambridge, 1996.

Allan, V., B. Backley, L. Felperin, N. James and H. Gee: *Sight and Sound Supplement*, BFI, London, 1996.

Astor, G.: *The "Last" Nazi: The Life and Times of Dr. Joseph Mengele*, Weidenfeld and Nicolson, London, 1985.

Ayala, F. J.: *The Genetic Structure of Populations*, W.H. Freeman & Co., San Francisco, California, 1977.

Bajema, C. J.: *Natural Selection in Human Populations, the Measurement of Ongoing Genetic Evolution in Contemporary Societies*, Wiley, New York, 1971.

Barnaby, W.: *Plague Makers: The Secret World of Biological Warfare*, Vision Paperbacks, London, 1999.

Bast, R.C. Jr.: *Cancer Medicine*, Decker, Hamilton, 2000.

Bauer, M.: *Resistance to New Technology: Nuclear Power, Information Technology and Biotechnology*, Cambridge University Press, Cambridge, England, 1995.

Berg, Paul, and Singer, Maxine: *Dealing with Genes: The Language of Heredity. Mill Valley*, University Science Books, CA, 1992.

Bernard, C.: *An Introduction to the Study of Experimental Medicine*, Dover Publications, New York, 1957.

Berry, R. J.: *Inheritance and Natural History*: Collins, London, 1977.

Bradley, W.G.: *Neurology in Clinical Practice*, Butterworth Heinemann, Boston, 2000.

Branagh, K., S. Lady, and F. Darabont: *Mary Shelley's Frankenstein: The Classic Tale of Terror Reborn*, Newmarket Press, London, 1994.

Brookes, Martin: *Get a Grip on Genetics*, *Time Life Books*, East Sussex, England, 1998.

Brosnan, J.: *The Primal Scream: A History of Science Fiction Film*, Orbit Books, London, 1991.

Bukatman, S.: *Blade Runner*, BFI, London, 1997.

Campbell, Neil A.; Brad Williamson; Robin J. Heyden: *Biology: Exploring Life*, Pearson Prentice Hall, Boston, Massachusetts, 2006.

Crick, F.: *Life Itself: Its Origin and Nature*, W.W. Norton, New York, 1982.

Curtis, M.: *The Geometry of DNA: A Structural Revision*, Blue Gallery, London, 1996.

Dams, R.D.: *Principles of Neurology*, McGraw Hill, New York, 1997.

Dawkins, R.: *The Selfish Gene*, Oxford University Press, New York, 1976.

Diaz, E.: *Microbial Biodegradation: Genomics and Molecular Biology*, Caister Academic Press, UK, 2008.

Dobzhanshy, T.: *Genetics of the Evolutionary Process*, Columbia University Press, New York, 1970.

Dulbecco, R.: *The Design of Life*, Yale University Press, New Haven, Connecticut, 1987.

Dunbar, Robert E.: *Heredity*, Franklin Watts Publisher, New York, 1978.

Edey, M. A., and Johanson, D. C.: *Blueprints: Solving the Mystery of Evolution*, Little, Brown and Co., Boston, Mass, 1989.

Fisher, R. A.: *The Genetical Theory of Natural Selection*, Clarendon Press, Oxford, 1930.

Fritz, A.: *International Classification of Diseases for Oncology*, World Health Organization, Geneva, 2000.

Gall, Joseph G.: *Landmark Papers in Cell Biology*, Cold Spring Harbor Laboratory Press, Plainview, NY, 2001.

Garza-Valdes, L.A.: *The DNA of God?* Hodder and Stoughton, London, 1998.

George S. Paul: *Beyond Humanity: Cyber Evolution and Future Minds*, Charles River Media, Roackland, 1996.

Gillis, Justin: *Drug Firms, Gene Labs to Map Genetic Code*, The Daily News, Longview, WA, 1999.

Glover, D. M., and Hames, B. D.: *Genes and Embryos*, Oxford University Press, New York, 1989.

Goldman, L.; Ausiello, D. A.: *Cecil Textbook of Medicine*, Saunders, Philadelphia, 2004.

Griffiths A.J.F.: *Introduction to Genetic Analysis*, W.H. Freeman and Company, New York, USA, 2005.

Halacy, D.S., Jr.: *Genetic Revolution, Shaping Life for Tomorrow*, Harper & Row, Publishers, New York, 1974.

Hamerton, J. L.: *Human Cytogenetics*, Academic Press, New York, 1971.

Hanley, R.: *Is Data Human? The Metaphysics of Star Trek*, Boxtree, London, 1998.

Hartl, Daniel L.: *Basic Genetics*, Jones and Bartlett Publishers, Boston, 1991.

Haubrich, W.S.: *Bockus Gastroenterology*, Saunders, Philadelphia, 1995.

Heider, J. and Rabus, R: *Microbial Biodegradation: Genomics and Molecular Biology*, Caister Academic Press, UK, 2008.

Holtz, Robert D.; William, Kovacs D.: *An Introduction to Geotechnical Engineering*, Prentice Hall, UK, 1981.

Jameson, J. L.: *Principles of Molecular Medicine*, Humana Press, Totowa, 1998.

Jonoska, N.: *Self-Assembling DNA Graphs, DNA-Based Computers VIII*, Springer-Verlag, Berlin, 2003.

Kelves, D.: *In the Name of Eugenics: Genetics and the Uses of Human Heredity*. Harmondsworth: Penguin.

Klein, Aaron E.: *Threads of Life: Genetics from Aristotle to DNA*, The Natural History Press, Garden City, New York, 1955.

Klug, William S. and Michael R.: Cummings. *Essentials of Genetics*, Prentice Hall, New Jersey, 1996.

Lakoff, G. and M. Johnson: *Metaphors We Live By*, University of Chicago Press, Chicago, 1980.

Lewin B.: *Genes VII*, Oxford University Press Inc., New York, USA, 2000.

Lewontin, R.C.: *The Doctrine of DNA: Biology as Ideology*, Penguin, Harmondsworth, 1993.

Lyon, Jeff and Gorner, Peter. Altered Fates: *Gene Therapy and the Retooling of Human Life*, W. W. Norton and Company, New York, 1995.

Margulis, L.: *Symbiosis in Cell Evolution,* W.H. Freeman, San Francisco, 1981.

Mayr, E.: *Change of Genetic Environment and Evolution*, Allen and Unwin, London, 1954.

Mayr, Ernst: *The Growth of Biological Thought: Diversity, Evolution, and Inheritance*, Harvard University Press, Cambridge, 2000.

Migloni, G.S.: *Dictionary of Plant Genetics and Molecular Biology*, Hawthorne Press, New York, 1998.

Nelkin, D. and M.S. Lindee: *The DNA Mystique: The Gene as a Cultural Icon*, W.H. Freeman, New York, 1995.

Nottingham, S.F.: *Eat Your Genes: How Genetically Modified Food is Entering Our Diet*, Zed Books, London, 1998.

Rietman, Ed.: *Molecular Engineering of Nanosystems,* Springer, New York, 2001.

Schummer, J.: *Interdisciplinary Issues in Nanoscale Research*, IOS Press, Amsterdam, 2004.

Scriver, C.R.: *The Metabolic and Molecular Basis of Inherited Disease*, McGraw Hill, New York, 2001.

Stebbins, G. L.: *Darwin to DNA, Molecules to Humanity*, W. H. Freeman, San Francisco, 1982.

Sturtevant, Alfred: *History of Genetics*, Harper and Row, New York, 1965.

Stwertka, Eve and Albert: *Genetic Engineering*, Franklin Watts, New York, 1982.

Suzuki, D., and Knudtson, P.: *Genethics: The Clash Between the New Genetics and Human Values*, Harvard University Press, Cambridge, Mass, 1989.

Watson, J. D.: *The Double Helix*, Antheneum, New York, 1968.

Williams, J.G. and R.K. Patient: *Genetic Engineering*, IRL Press, Oxford, England, 1988.

Wilson, E. O., and Lumden, C.: *Genes, Mind, and Culture: The Evolutionary Process*, Harvard University Press, Cambridge, Mass, 1981.

Wilson, E. O.: *Biophilia*, Harvard University Press, Cambridge, Mass, 1985.

Winchester, A. M.: *Heredity, Evolution and Humankind*, West Publishing Co., St. Paul, Minn., 1976.

Winston, R.: *The Future of Genetic Manipulation*, Phoenix, London, 1997.

Wright, L.: *Twins: Genes, Environment and the Mystery of Human Identity*, Weidenfeld & Nicolson, London, 1997.

Zimmerman, E. G.: *Karyology, Systematics, and Chromosomal Evolution in the Rodent Genus, Sigmodon*, Michigan State Univ., UK 1970.

Index

A

Amino Acid, 13, 34, 36, 37, 42, 59, 64, 65, 66, 67, 68, 181, 183, 187, 188, 221, 222.

B

Base Pair, 24, 25, 29, 30, 31, 33, 62, 67.
Breeding, 6, 12, 108, 110, 129, 134, 135, 158, 159, 164, 166, 173, 176, 177, 184, 189, 201, 202, 203, 204, 205, 206, 207, 208, 209, 227, 229, 230, 231, 233, 238, 240, 243, 247, 249.

C

Cancer, 12, 46, 48, 49, 196, 199, 200, 201, 217, 219, 220, 221.
Cell, 5, 6, 7, 15, 16, 17, 19, 21, 22, 23, 24, 26, 29, 36, 41, 42, 43, 44, 45, 47, 48, 50, 55, 58, 59, 61, 64, 65, 66, 67, 68, 70, 71, 84, 87, 89, 114, 144, 163, 181, 195, 196, 197, 198, 241, 242, 246, 256, 257, 264, 270, 272.
Cell Cycle, 49, 152.
Cell Nucleus, 16, 36.
Chemotherapy, 49.
Chromatin, 39, 40, 41, 43, 45, 48, 50, 54, 55, 59, 60, 61, 62, 63, 65.

Chromatin Remodeling, 43, 45, 62, 65.
Chromosome, 1, 2, 3, 5, 6, 7, 9, 15, 16, 18, 22, 33, 36, 39, 40, 46, 47, 49, 53, 55, 56, 81, 91, 101, 124, 160, 161, 163, 164, 165, 166, 168, 170, 171, 173, 174, 175, 176, 177, 178, 179, 180, 189, 190, 195, 196, 204, 212, 224, 225, 227, 240, 252, 264.
Circuitry, 53.
Cloning, 39, 58, 180, 181, 182, 183, 185, 186, 188, 190, 191, 233, 234, 252, 260, 261.
Contractive System, 70.
Cytogenetices, 80.

D

Deoxyribonucleic Acid, 34.
Diploid Chromosome, 91, 124.
Disease, 6, 12, 14, 20, 44, 45, 47, 48, 80, 84, 147, 153, 160, 162, 173, 185, 187, 195, 196, 198, 199, 215, 218, 219, 220, 222, 223, 224, 225, 226, 227, 229, 243.
Diversity, 11, 17, 61, 65, 184, 194, 202, 203, 204, 228, 229, 230, 231, 232, 233, 235, 237, 239, 247, 250, 268.

Drosophila, 52, 53, 59, 63, 64, 135, 170, 185, 215, 228, 246.

E

Enzyme, 3, 13, 24, 26, 40, 41, 42, 52, 57, 181, 184, 200, 221, 223, 224, 225.
Ethanol, 154, 155.
Eukaryotes, 16, 37, 50, 52, 54, 57, 61, 63, 64, 248.
Eukaryotic Cells, 18, 53, 54, 55, 248.

F

Fluorophores, 28.

G

Gene, 3, 4, 5, 6, 11, 12, 13, 14, 15, 17, 18, 20, 23, 24, 36, 37, 38, 39, 40, 42, 43, 45, 46, 47, 48, 49, 50, 51, 52, 53, 54, 55, 56, 57, 58, 59, 60, 61, 62, 63, 64, 65, 66, 85, 88, 108, 110, 112, 113, 125, 161, 162, 164, 170, 171, 172, 177, 178, 179, 180, 181, 182, 183, 185, 186, 187, 188, 189, 190, 191, 192, 195, 196, 198, 199, 200, 203, 205, 206, 207, 211, 212, 213, 214, 215, 216, 217, 218, 219, 220, 222, 223, 225, 227, 229, 231, 232, 233, 234, 235, 236, 237, 238, 239, 241, 242, 244, 245, 246, 247, 248, 249, 252, 256, 257, 260, 261, 262, 263, 265, 268.
Gene Expression, 37, 39, 40, 43, 45, 49, 50, 51, 52, 53, 54, 56, 58, 59, 60, 181, 182, 215.

Gene Interaction, 12.
Gene Therapy, 217, 223.
Genetic Architecture, 225.
Genetic Code, 34, 38, 65, 66, 67, 68.
Genetic Engineering, 181, 184, 207, 239, 240.
Genetic Hitchhiking, 234.
Genetic Linkage, 1, 2, 3, 4, 7, 188, 189.
Genetic Pollution, 233, 238, 239, 240, 241.
Genetic Testing, 225.
Genetic Variation, 14, 158, 195, 196, 197, 226, 229, 235, 244, 247, 248.
Genetics, 1, 5, 7, 10, 11, 12, 15, 16, 24, 37, 38, 48, 55, 80, 111, 130, 197, 201, 210, 211, 212, 214, 215, 216, 217, 218, 219, 220, 225, 226, 227, 228, 241, 242, 243, 244, 249, 250, 252, 253, 269.
Genome, 15, 16, 17, 18, 24, 25, 33, 37, 38, 40, 42, 44, 45, 50, 56, 58, 62, 65, 135, 160, 162, 164, 168, 170, 171, 181, 190, 191, 195, 196, 206, 210, 214, 215, 216, 217, 218, 221, 225, 230, 245, 246, 253, 261, 262, 265.

H

Helix, 26, 28, 34, 35, 51, 57, 64, 68, 70, 72, 74.
Herbicides, 203, 207.
Heredity, 1, 7, 10, 19, 38, 244.
Hormone, 53, 59, 200, 255, 258, 261.
Human Genome, 16, 17, 24, 56, 195, 196, 215, 216, 225, 265.

I

Inhibitors, 55.

K

Karyotype, 18, 80, 81, 221.
Kinetics, 28.

M

Mapping, 2, 3, 50, 102, 103, 164,
    166, 168, 170, 171, 172,
    173, 180, 188, 189, 214,
    215, 243, 255.
Mechanism, 10, 11, 24, 40, 45,
    69, 73, 75, 76, 77, 78, 80,
    223, 260.
Medical Genetics, 217, 218, 219,
    225.
Medicine, 26, 28, 194, 195, 217,
    218, 248.
Mendelian Inheritance, 10, 210.
Metal Base Pairs, 31.
Metaphase, 81, 164, 173.
Methylation, 39, 40, 41, 42, 43,
    45, 48, 49, 50, 51, 59, 135,
    169, 223.
Micropropagation, 130, 132, 136,
    138, 147, 148, 149, 150,
    151, 154.
Microscopy, 72, 218.
Mitosis, 81, 173.
Molecular Biology, 15, 24, 26, 66,
    153, 205, 214, 216.
Molecular Genetics, 10, 211, 215,
    243, 244.
Molecular Tools, 253.
Myofilament, 71.
Myofilaments, 68, 69, 70, 72, 74.

N

Nervous System, 59, 251.
Nucleic Acid, 19, 26, 27, 29, 36,
    65.

Nucleotide, 17, 25, 28, 34, 35,
    36, 64, 66, 195, 196, 215,
    235, 236.

O

Orthogonal System, 32.
Oxidative Phosphorylation, 59,
    76.

P

Panmixis, 83, 84, 85, 88, 89, 90,
    92, 94, 95, 98, 100, 103,
    121, 122.
Pesticides, 152, 153.
Plant Breeding, 108, 158, 202,
    203, 205, 207, 208, 209.
Plant Genetic, 189.
Plant Tissue, 148.
Plant tissue, 148, 149.
Prokaryotes, 37, 46, 51, 52, 61,
    248.
Protein, 21, 22, 23, 24, 28, 30,
    34, 36, 37, 41, 44, 49, 50,
    51, 52, 53, 54, 55, 56, 57,
    62, 63, 64, 65, 66, 67, 68,
    69, 70, 71, 72, 79, 162,
    181, 182, 183, 188, 200,
    207, 214, 215, 222, 245,
    246, 252.
Protein Synthesis, 34, 36, 57, 65,
    67, 68.

R

Random Mating, 84, 85, 96, 101,
    108, 113, 123, 242.
Ribonucleic Acid, 36.

S

Sacromere I-band, 69.
Sarcoplasmic Reticulum, 77, 78, 79,
    80.
Sequence, 16, 17, 18, 27, 28, 30,
    32, 33, 34, 36, 37, 38, 41,

42, 43, 45, 46, 50, 51, 52, 59, 61, 63, 64, 65, 66, 67, 74, 75, 134, 163, 164, 170, 179, 181, 182, 183, 185, 186, 188, 189, 190, 191, 195, 200, 206, 214, 215, 216, 222, 245, 263.

Sequencing, 16, 17, 26, 27, 39, 50, 57, 181, 214, 215, 216, 222.

Sex Determination, 53.

Sex-linked Inheritance, 9.

Slioing Mechanism, 73.

Syndrome, 45, 46, 218, 219, 222, 225.

T

Tissue, 48, 53, 129, 130, 131, 132, 133, 134, 135, 136, 137, 140, 141, 147, 148, 149, 150, 151, 153, 156, 159, 161, 181, 182, 203, 204, 214, 215, 217, 233.

Troponin, 72, 75, 76, 80.

X

X Chromosome, 9, 39, 81.

x-ray, 74, 75.

Y

Y Chromosome, 9, 177.

❑❑❑